# TREVOR KLETZ COMPENDIUM

# TREVOR KLETZ COMPENDIUM
## His Process Safety Wisdom Updated for a New Generation

ANDY BRAZIER
*Risk Consultant, AB Risk Limited, Llandudno, United Kingdom*

DAVID EDWARDS
*Visiting Professor of Safety and Loss Prevention, School of Aeronautical, Automotive, Chemical and Materials Engineering, Loughborough University, United Kingdom*

FIONA MACLEOD
*Managing Director, Billions Europe Ltd., Stockton-on-Tees, United Kingdom*

CRAIG SKINNER
*Operational Safety Director, BP, Surrey, United Kingdom*

IVAN VINCE
*Risk Consultant, ASK Consultants, Beckenham, United Kingdom*

ELSEVIER

Elsevier
Radarweg 29, PO Box 211, 1000 AE Amsterdam, Netherlands
The Boulevard, Langford Lane, Kidlington, Oxford OX5 1GB, United Kingdom
50 Hampshire Street, 5th Floor, Cambridge, MA 02139, United States

**Notices**
Knowledge and best practice in this field are constantly changing. As new research and experience broaden our
understanding, changes in research methods, professional practices, or medical treatment may become
necessary.

Practitioners and researchers must always rely on their own experience and knowledge in evaluating and using
any information, methods, compounds, or experiments described herein. In using such information or methods
they should be mindful of their own safety and the safety of others, including parties for whom they have a
professional responsibility.

To the fullest extent of the law, neither the Publisher nor the authors, contributors, or editors, assume any liability
for any injury and/or damage to persons or property as a matter of products liability, negligence or otherwise, or
from any use or operation of any methods, products, instructions, or ideas contained in the material herein.

**Library of Congress Cataloging-in-Publication Data**
A catalog record for this book is available from the Library of Congress

**British Library Cataloguing-in-Publication Data**
A catalogue record for this book is available from the British Library

ISBN: 978-0-12-819447-8

For information on all Elsevier publications
visit our website at https://www.elsevier.com/books-and-journals

*Publisher:* Susan Dennis
*Acquisitions Editor:* Kostas KI Marinakis
*Editorial Project Manager:* Andrea Dulberger
*Production Project Manager:* Kumar Anbazhagan
*Cover Designer:* Victoria Pearson

Typeset by SPi Global, India

Working together
to grow libraries in
developing countries

www.elsevier.com • www.bookaid.org

# Contents

### Introduction

1 Trevor Kletz  1
2 Useful links  5
3 About this book  8
References  8

### 1. Hazard and operability (HAZOP) analysis

1.1 Introduction  9
1.2 HAZOP organisation and process  20
1.3 Limitations of HAZOP  30
1.4 Practical application  38
References  44

### 2. Hazan (Quantified risk assessment)

2.1 Introduction  47
2.2 Risk criteria  51
2.3 Frequency/probability assessment  56
2.4 Consequence analysis  66
2.5 Pitfalls in Hazan  68
References  71

### 3. Inherent safety

3.1 Introduction  73
3.2 The concept of inherent safety  78
3.3 Principles of inherent safety  85
3.4 Applying inherent safety through a system's lifecycle  97
3.5 How far should we go?  106
References  113

### 4. Managing maintenance risk

4.1 Introduction  115
4.2 Identifying maintenance requirements  119
4.3 Preparing for maintenance  125
4.4 Safety during maintenance  144
4.5 Conclusions  167
References  168

### 5. Control of modifications

5.1 Introduction  171
5.2 Different types of change  172
5.3 Modifications made during maintenance  174
5.4 Process modifications  181
5.5 Modifying written procedures  183
5.6 Software modifications  185
5.7 Changes in organisation  186
5.8 Gradual changes  190
5.9 Modifications to improve the environment  192
5.10 Managing the risks  193
References  200

### 6. Human error

6.1 Introduction  203
6.2 Different types of error  204
6.3 Reducing human error probability  224
6.4 Human factors analysis  240
References  244

### 7. Accident investigations—Missed opportunities

7.0 Introduction  247
7.1 Accident investigations often find only a single cause  249
7.2 Accident investigations are often superficial  252
7.3 Accident investigations list human error as a cause  254
7.4 Accident reports look for people to blame  256
7.5 Accident reports list causes that are difficult or impossible to remove  258
7.6 We change procedures rather than designs  261
7.7 We may go too far  263
7.8 We do not let others learn from our experience  264

7.9 We read or receive only overviews  266
7.10 We forget the lessons learned and allow the
accident to happen again  268
7.11 We examine each accident in isolation  271
References  272

**Appendix  275**
A calendar of disasters  275
1 January: Feyzin  276
2 February: Georgia Sugar  277
3 March: Texas City  278
4 April: Chernobyl  279

5 May: Wanggongchang Armoury Explosion  281
6 June: Flixborough  283
7 Chevron Pembroke Explosion  284
8 July: Piper Alpha  286
9 August: Banqiao  288
10 September: Longford  290
11 October: Phillips Pasadena  292
12 November: Sandoz  294
13 December: Bhopal  295
14 Buncefield  297
References  298

**Index  299**

# Introduction

## 1 Trevor Kletz

One of the founders and leaders of process safety in thought and practice, Professor Trevor Kletz died on 31 October 2013. He left behind a magnificent canon of publications, including 16 books and well over a 100 reviewed papers on loss prevention and process safety, which communicate useful learning in a readable and understandable way and which will serve long into the future to guide safety people in their work in our industry and beyond. They have provided the philosophical and practical basis for many areas of process safety and encouraged people to look at safety differently.

Kletz wrote with clarity and understood the power of story as a way to get important messages across. The aim of this compendium is to update and combine Kletz's writings with new stories and commentary showing how they remain relevant today and how the underlying ideas have been developed further. It is a resource for a new generation of engineers and others, which is intended to inspire them to make a positive contribution to the way safety is managed where they work. It should also prompt people who have read his books in the past to have another look.

Trevor Asher Kletz was born in 1922 in Darlington of Jewish parents, from a Russian immigrant background. His father, a shopkeeper, was insistent that Trevor should better himself and he attended The King's School, Chester and then Liverpool University. When he was 11 years old, an uncle had given him a chemistry set as a present, which influenced his decision to study chemistry. He graduated in 1944 and joined Imperial Chemical Industries (ICI), where he spent 8 years in research, 16 in production management, and the last 14 as safety adviser to the Petrochemicals Division. In 1978 he was appointed an industrial professor in the Chemical Engineering Department at Loughborough University. On retiring from ICI in 1982 he joined the Department full-time; in 1986 he became a visiting fellow and was latterly a visiting professor at Loughborough and an adjunct professor at Texas A&M University in the United States.

David Edwards became aware of Trevor when he joined the Loughborough Chemical Engineering Department in 1990. He recalls: *"I did not know much about how academic careers progressed but one of the few things they told me was to get a line of research. Safety seemed the most prominent of the lines available (mainly due to Frank Lees), so I aligned myself with that. Then I attended one of Trevor's lectures on inherently safer design, which blew me away - isn't all plant designed like this (it's just common sense) - well, actually, no. One thing that particularly struck me was Trevor's graph showing that safety improves with money expended. We spent the next few years working, with considerable guidance from Trevor, to show that inherently safer plants are also cost effective."*

Trevor was appointed an OBE in 1997 and he was a Fellow of the Royal Academy of Engineering, the Institution of Chemical Engineers, the Royal Society of Chemistry, and the American Institute of Chemical Engineers. He was also an Honorary Fellow of the Institute of Occupational Safety and Health and the Safety and Reliability Society. He was one of the most famous chemical engineers, who was not a chemical engineer!

Trevor retired at the age of 90, as his old school noted: *"One of King's oldest Old Boys Trevor Kletz has finally retired - what a career! Every engineer/technician in the UK and far beyond on any chemical plant will have heard of Trevor, who has much improved the safety of the chemical process industry with his career work. He certainly deserves his place in the School's Hall of Fame, and all the accolades he has been awarded in his long career - Happy retirement Trevor!"*

Up until then he was still making forthright and insightful statements about safety in the process industry, saying in 2011 that the industry's 'macho culture' was one of the main causes of recent accidents.

Jill Wilday, who knew him at ICI, commented: *"When I joined ICI in the late 70s he was the Safety Advisor for Petrochemicals Division and his safety newsletter was circulated throughout ICI and was as popular as New Scientist among young engineers. (This was well before email as a means of sharing information). The newsletter predominantly described incidents and their recommendations, but always with an opportunity for you to work out what went wrong before reading as far as the conclusions of the internal enquiry. It was easy and entertaining to read and consequently it was popular, well read, and so it acted to improve everyone's understanding. He also used humour and those bits in particular were much repeated in conversations in the canteen."*

Andy Rushton was a colleague of Trevor and David Edwards in the Chemical Engineering Department at Loughborough University, where they worked together on inherent safety. Andy says: *"He was a great communicator (always top of the poll on the [student feedback] 'happy' sheets) and his forte was distilling his (and others') experience, drawing out principles and presenting them in a relevant way (not just to process engineers – transport, defence and other sectors have been mightily influenced too). He could make what he was saying relevant to you and your problems, and could present it with humour, patience and cogency (although not everyone agreed with everything he said)."*

His entertaining after-dinner stories further served to make people remember him and his messages.

Not only was he a masterful communicator, both written and orally, but he also had the insight to reduce seemingly complicated issues to the simple fundamentals and to understand which were important. He knew that he could save lives by spreading his insights and he had the perseverance, patience, and generosity to repeat his messages until heard and understood.

So what were his messages? Well, if you don't know, you should definitely read this compendium! We also recommend that you read his books, which are listed here.

| Title | First published | Latest edition | Publisher |
| --- | --- | --- | --- |
| Cheaper, safer plants | 1984 | 1984 | IChemE |
| Myths of the chemical industry | *1984* | *1984* | IChemE |
| An Engineer's view of human error | 1985 | 3rd—2001 | IChemE |
| What went wrong? | 1985 | 6th—2019 | Elsevier |
| Hazop and Hazan—Notes on the identification of hazards | 1986 | 1986 | IChemE |
| Learning from Accidents | 1988 | 3rd—2001 | Elsevier |
| Critical aspects of safety and loss prevention | 1990 | 1st—1990 | Elsevier |
| Dispelling chemical engineering myths | 1990 | 3rd—1996 | CRC Press |
| Improving chemical engineering practices | *1990* | *2nd—1990* | CRC Press |
| Plant design for safety | 1991 | 1st—1991 | CRC Press |
| Hazop and Hazan | 1992 | 4th—1999 | IChemE |
| Lessons from disaster | 1993 | 1st—1993 | IChemE |
| Computer control and human error | 1995 | 1st—1995 | IChemE |
| Process Plants—a handbook for inherently safer design | 1998 | 2nd—2010 | CRC Press |
| By accident (autobiography) | 2000 | 1st—2000 | PFV Publications |
| Still going Wrong | 2003 | 1st—2003 | Elsevier |

If you only read one of these, choose "Process Plants: A Handbook for Inherently Safer Design" [1], now in its second edition and co-authored by Professor Paul Amyotte of Dalhousie University. The inherently safer approach aims to eliminate or reduce hazards or exposure to them or the chance of occurrence by design. Most people will say that it is common sense, and it is, but it took Trevor to cast this common sense into a practical philosophy. There are undoubtedly tens, probably hundreds, possibly thousands of people who each day go home to their families and will do for many years to come, but who are only with us because of decisions influenced by this light that Trevor shone to lead our way.

Craig Skinner, also a chemical engineering graduate of Loughborough University, remarks: "*It wasn't until I was a practicing chemical engineer that Kletz's inherent safety really started to make sense to me, and the simple and practical framework of inherent safety principles has helped in many key engineering decisions over the last 30 years, both when applied to major hazards and when deciding between engineering options. Over this period, inherent safety has also grown from a philosophy into a practical, effective process and team activity, and this is shared in this book alongside Kletz's original concepts and stories, which we hope readers find pragmatic and helpful.*"

Inherent Safety is not the only area where his clear thinking has changed the way we think and act. His writings on human error and accident investigation refocused the emphasis away from individual lapses to systems failures and safer design. These concepts fostered a revolution in modern safety management thinking. In a video that he made for the U.S. Chemical Safety Board, Trevor says: *"For a long time, people were saying that most accidents were due to human error and this is true in a sense but it's not very helpful. It's a bit like saying that falls are due to gravity."* Andy Brazier comments: *"The discovery of Kletz's book 'An Engineer's view of human error'* [2] *was a defining moment for me because it gave a clear insight into the issues affecting industry and illustrated why understanding human factors is so important."*

A theme that runs through Trevor's work is drawing lessons from accidents and his mantra: *"organisations have no memory"* should be a constant watchword. Another favourite among his many sayings is: *"There's an old saying that if you think safety is expensive, try an accident. Accidents cost a lot of money. And, not only in damage to plant and in claims for injury, but also in the loss of the company's reputation."*

Trevor also said: *"You may not agree with some (or even all) of the advice in my books but I hope you won't disregard the accident reports. If you don't like my advice, I hope you will decide what to do instead."* [3].

Trevor was a firm believer that people should be persuaded by sound reason to take the safer course. His was a common sense approach and if we put all of his ideas together we get a common sense philosophy.

1. **Inherent Safety**: eliminate the hazard or cause of the accident.
2. If you can't do this, use techniques such as **HAZOP and HAZAN** to analyse what you want to do, applying rigorous technical methods but also with an awareness of **Human Error**, in order to minimise the risk.
3. If despite all of this, you do have an accident: investigate it, record, and make the details easily accessible, **learn the lessons** and **tell people** all about it, so that we don't do it again.

Outside of his professional life and vocation, Trevor was very active in the Jewish community and had a strong interest in steam trains. He also lived in a bungalow. This is inherently safer, because the hazards due to stairs, which are the biggest cause of accidents in the home, have been eliminated. He always wore belt and braces, so that there were two layers of protection against his trousers falling down; he probably had a piece of string in his pocket just in case!

Trevor is buried in the Jewish section of the Linthorpe Road Cemetery in Middlesbrough, surrounded by trees and birdsong.

The late Professor Sam Mannan, who was Director of the Mary Kay O'Connor Process Safety Center at Texas A&M University, summed up his life and work well: *"Some have characterized Trevor as a scholar, some have called him an astute practitioner, and some hold him in high regard for his unique ability to transform complex issues into simple messages that he communicated in his unique way. Above all, Trevor was a visionary and a trailblazer, the likes of whom come in our midst only every few centuries."*

Managing process safety is proving to be a long-term challenge for industry. Unfortunately major accidents are still occurring and lessons from the past are not being learnt as well as they should. Trevor was one of the first people to tackle these issues and his stories and proposals for improving safety remain entirely relevant today. This compendium collects, updates, and enhances these messages for a new audience.

David Edwards said "It is a great honour for me to contribute to this new compendium of Trevor's work. He was a great man, whom I had the privilege of working with and calling my friend and mentor. It behoves us all to honour his memory by following his teaching and example in our professional and personal lives by learning from past incidents and making all of our endeavours inherently safer."

## 2 Useful links

As well reading Kletz's books, we encourage you to use the many resources available via the internet to give you a deeper understanding of the topics covered in this book. You will find the following to be particularly useful.

### 2.1 U.S. Chemical Safety and Hazard Investigation Board (CSB)—https://www.csb.gov/

This organisation investigates industrial chemical accidents that occur in the United States. They have published numerous reports, but it is their videos that are particularly valuable. The following reports and videos are particularly relevant to the content of this book (all free to access):

- BP Texas City refinery—https://www.csb.gov/bp-america-refinery-explosion/
- Formosa Plastics Vinyl Chloride Explosion—https://www.csb.gov/formosa-plastics-vinyl-chloride-explosion/
- Millard Refrigerated Services Ammonia Release—https://www.csb.gov/millard-refrigerated-services-ammonia-release/
- Husky Energy Refinery Explosion and Fire—https://www.csb.gov/husky-energy-refinery-explosion-and-fire/
- Chevron Refinery Fire—https://www.csb.gov/chevron-refinery-fire/
- Sterigenics Ethylene Oxide Explosion—https://www.csb.gov/sterigenics-ethylene-oxide-explosion/
- Caribbean Petroleum Refining Tank Explosion and Fire—https://www.csb.gov/caribbean-petroleum-refining-tank-explosion-and-fire/
- CSB Video Excerpts from Dr. Trevor Kletz—https://www.csb.gov/videos/csb-video-excerpts-from-dr-trevor-kletz/

### 2.2 UK Health and Safety Executive (HSE)—https://www.hse.gov.uk/

The HSE regulates and enforces health and safety in the United Kingdom. It publishes a great deal of guidance and some accident reports. The following are particularly relevant to the content of this book (all free to access):

- HSG 250 Guidance on permit-to-work systems—https://www.hse.gov.uk/pubns/books/hsg250.htm
- HSG 253 The safe isolation of plant and equipment—https://www.hse.gov.uk/pubns/books/hsg253.htm

- Human factors: Inspectors human factors toolkit—https://www.hse.gov.uk/humanfactors/toolkit.htm
- Competent Authority procedures and delivery guides—https://www.hse.gov.uk/comah/ca-guides.htm
- Buncefield reports and recommendations—https://www.hse.gov.uk/comah/buncefield/index.htm
- Chevron Pembroke Amine regeneration unit explosion—https://www.hse.gov.uk/comah/chevron-pembroke-report-2020.pdf

## 2.3 Institution of Chemical Engineers (IChemE)—https://www.icheme.org/

The IChemE is the professional body for Chemical Engineers in more than 100 countries. It is very active in the field of process safety. The following sources of information are particularly relevant to the contents of this book (some free and some chargeable):

- Loss Prevention Bulletin—https://www.icheme.org/knowledge/loss-prevention-bulletin/
- Hazards Conference proceedings—https://www.icheme.org/membership/communities/special-interest-groups/safety-and-loss-prevention/resources/hazards-archive/
- Archive of ICI newsletters, mainly produced by Trevor Kletz—https://www.icheme.org/membership/communities/special-interest-groups/safety-and-loss-prevention/resources/ici-newsletters/
- HSE accident reports (collection of out-of-print reports)—https://www.icheme.org/membership/communities/special-interest-groups/safety-and-loss-prevention/resources/hse-accident-reports/
- The IChemE Safety Centre is a not-for-profit multi-company, subscription based, industry consortium, focused on improving process safety—https://www.icheme.org/knowledge/safety-centre/

## 2.4 Energy Institute (EI)—https://www.energyinst.org/

EI is a chartered membership organisation for professionals working in the energy industries. It has a number of working parties focused on process safety issues and has published a number of guidance documents. The following sources of information are particularly relevant to the contents of this book (some free and some chargeable):

- Human factors safety critical task analysis—https://publishing.energyinst.org/topics/human-and-organisational-factors/guidance-on-human-factors-safety-critical-task-analysis2
- Applying inherent safety in design—https://publishing.energyinst.org/topics/process-safety/guidance-on-applying-inherent-safety-in-design-reducing-process-safety-hazards-whilst-optimising-capex-and-opex

- Bow ties in risk management—https://publishing.energyinst.org/topics/process-safety/risk-management/bow-ties-in-risk-management-a-concept-book-for-process-safety
- Human factors briefing notes—https://www.eemua.org/Products/Publications/Digital/EEMUA-IIS2-Cyber-security.aspx

## 2.5 Engineering Equipment and Materials Users Association (EEMUA)—https://www.eemua.org

EEMUA is a membership organisation that publishes guidance and provides training on a range of technical engineering topics. The following sources of information are particularly relevant to the contents of this book (mostly chargeable):

- EEMUA 191 Alarm management—https://www.eemua.org/Products/Publications/Digital/EEMUA-Publication-191.aspx
- EEMUA 201 Control room design—https://www.eemua.org/Products/Publications/Digital/EEMUA-Publication-201.aspx
- EEMUA 222 Application of IEC 61511 to safety instrumented systems—https://www.eemua.org/Products/Publications/Digital/EEMUA-Publication-222.aspx
- Cyber security—https://www.eemua.org/EEMUAPortalSite/media/EEMUA-Flyers/EEMUA-Industry-Information-Sheet-2.pdf

## 2.6 International Association of Oil & Gas Producers (IOGP)—https://www.iogp.org/

IOGP acts as the voice of the global upstream oil and gas industry. It shares knowledge and good practices to achieve improvements in health, safety, the environment, security, and social responsibility. The following sources of information are particularly relevant to the contents of this book (some free and some chargeable):

- IOGP 456 Process Safety Key Performance Indicators—https://www.iogp.org/bookstore/product/process-safety-recommended-practice-on-key-performance-indicators/
- IOGP 454 Human factors engineering in projects—https://www.iogp.org/bookstore/product/human-factors-engineering-in-projects/

## 2.7 Mary Kay O'Connor Process Safety Center—http://psc.tamu.edu/

The Mary Kay O'Connor Process Safety Center was established in 1995 in memory of Mary Kay O'Connor, an Operations Superintendent killed in an explosion on 23 October 1989 at the Phillips Petroleum Complex in Pasadena. Its mission is to lead the integration of process safety through education, research, and service into learning and practice of all individuals and organisations. It publishes a range of white papers, position statements, and technical papers, and provides links to other useful sources of information.

Trevor was associated with the Centre for many years, with his own 'Trevor's Corner', which published short topical articles on safety and related matters—even the dangers of Powerpoint http://psc.tamu.edu/resources/trevors-corner/trevors-corner-archives.

## 2.8 Center for Chemical Process Safety (CCPS)—https://www.aiche.org/ccps

CCPS is part of the American Institute of Chemical Engineers (AIChemE). It aims to lead the way in improving industrial process safety by defining and developing useful, time-tested guidelines that have practical application within industry. The following sources of information are particularly relevant to the contents of this book (mostly chargeable):

- Inherently Safer Chemical Processes—https://www.aiche.org/ccps/resources/publications/books/guidelines-inherently-safer-chemical-processes-life-cycle-approach-3rd-edition
- Investigating Process Safety Incidents—https://www.aiche.org/ccps/resources/publications/books/guidelines-investigating-process-safety-incidents-3rd-edition
- Management of Change for Process Safety—https://www.aiche.org/ccps/publications/books/guidelines-management-change-process-safety
- Initiating events and independent protection layers in layer of protection analysis—https://www.aiche.org/ccps/resources/publications/books/guidelines-initiating-events-and-independent-protection-layers-layer-protection-analysis

## 2.9 European Process Safety Centre (EPSC)—https://epsc.be/

EPSC organises member meetings across Europe in order to plan and implement work on topics which are of most value to members. These can result in a tangible output such as information sheets, internal reports, publications, and international conferences. A monthly learning sheet is available as a free download—https://epsc.be/Learning+Sheets.html.

## 3 About this book

The authors have worked together to create the content of this compendium with support from the Institution of Chemical Engineers (IChemE) and with the approval of Kletz's family. Andy Brazier has taken the lead role in organising this endeavour both in terms of generating content, coordinating author input, and liaising with IChemE and Elsevier.

## References

[1] T. Kletz, P. Amyotte, Process Plant, A Handbook for Inherently Safer Design, second ed., Taylor & Francis, 2010.
[2] T. Kletz, An Engineer's View of Human Error, third ed., IChemE, 2001 First published 1985.
[3] T. Kletz, What Went Wrong? fifth ed., Elsevier, 2009.

# Hazard and operability (HAZOP) analysis

## 1.1 Introduction

Process Hazard Analysis (PHA) is considered to be one of the key methods when managing process safety. Hazard and Operability (HAZOP) is the most well-known, widely used, and effective PHA method.

Kletz stated in his autobiography [1] that there was an "obvious need to find out what can go wrong without waiting until it has gone wrong. HAZOP is the preferred technique in the process industries."

Kletz has sometimes been credited with developing the concept of HAZOP, but that is not true. HAZOP grew out of 'critical examination', a technique which was popular with ICI[a] during the 1960s for examining management decisions. Ken Gee, a production manager with ICI, decided to apply this technique to the design of a new phenol plant and, over a 4-month period, spent 3 days a week, every week, examining every aspect of the plant, discovering many potential hazards and operating problems that would not have been foreseen otherwise. Kletz said this was the first recognisable HAZOP [1].

Kletz used the following quote to illustrate the role of HAZOP for predicting potential problems [2]:

---

**Text Box 1.1**

Samuel Coleridge described history as a 'lantern on the stern', illuminating the hazards the ship has passed through rather than those that lie ahead. It is better to illuminate the hazards we have passed through than not illuminate them at all as we may pass the same way again, but we should try to see them before we meet them. HAZOP can be a lantern on the bow.

---

[a] ICI was a large British chemicals company, which is now defunct, but parts of it are now incorporated into other companies and many of the old sites are still occupied by process plants.

Although Kletz did not develop the method, he had a significant role in its adoption across the process industry. He encouraged public courses to be held at UK universities [3], with the first at Teesside Polytechnic (now a University) in 1975 and followed by UMIST in 1978. Kletz's motivation was the knowledge that HAZOP led to safer process plants and operations, but it is a sign of his communication skills that he not only persuaded universities to run courses but he was also successful at obtaining agreement from his employer, ICI, to make the method widely available to others, including competitors.

HAZOP has become the method of choice for identifying hazards and operability issues in process plant design. In most HAZOPs more operability problems are identified than hazards. Like all methods it has its limitations. However, HAZOP is used widely beyond the process industry, where the approaches taken are adapted to suit the particular requirements.

## 1.1.1 What is HAZOP?

Kletz described HAZOP as follows [2].

---

**Text Box 1.2**

HAZOP is a technique which provides opportunities for people to let their imaginations go free and think of all possible ways in which hazards or operating problems might arise, but—to reduce the chance that something is missed—it is done in a systematic way, and each pipeline and each sort of hazard is considered in turn. The study is carried out by a team so that the members can stimulate each other and build upon each other's ideas.

---

A more formal definition is provided by Eggett and Whitty [4]:

HAZOP is a formal, qualitative, systematic and rigorous examination of a plant, process or operation, in order to identify credible deviations from the design intent in the context of the complete system, which can contribute to the realisation of hazards or operability problems, by applying the experience, judgement and imagination, stimulated by key words, of a team.

The core principle of HAZOP is that problems only arise when there is a deviation from either normal operation or a system's design intent [4].

### 1.1.1.1 HAZOP basics

HAZOP is a qualitative process involving a structured and systematic examination of a system. The first activity is to break the system into 'nodes', each of which has a clear identity but is not too large to assess. For process plant the current approach is to identify nodes by section or by major equipment item. The main working document for a HAZOP is the Piping and Instrumentation Diagram (P&ID).

The objectives of HAZOP are to identify all the deviations that can have a consequence and decide whether action is required to control the hazard or operability issue. This is performed by the assessment team in a meeting or workshop, which is recorded by a scribe.

Documenting the HAZOP provides an effective means of capturing inputs and outputs and communicating potential hazards and operability problems to the people who operate, maintain, and manage the facility.

### 1.1.1.2 A simple example

A simple node consisting of a tank and associated pumps and valves is shown in Fig. 1.1 as a very simple example. The design intent is stated as follows: 'Pump P1 supplies petrol to a storage tank, which is vented to a safe location. P2 delivers petrol from the tank'.

During the HAZOP the node is examined and the scribe records the output of the team's deliberations and conclusions. This information is recorded as shown in Table 1.1.

This is only a very simple example. A full HAZOP would consider many other deviations and the plant would have other nodes.

### 1.1.1.3 Parameters and guidewords

A HAZOP is conducted by considering a range of parameters applicable to the type of system and a range of guidewords applicable to the parameter. The aim is to assist the HAZOP team to consider all possible deviations. This helps them to avoid becoming fixated on the more obvious or known issues, which may mean others that may be less obvious but potentially significant are overlooked.

In the process industry the typical parameters include flow, pressure, and temperature, and the guidewords include none, more, less, and different. They have been developed to

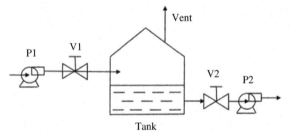

FIG. 1.1    Example node for a simple HAZOP.

TABLE 1.1    Example of the information captured in a HAZOP.

| # | Deviation | Cause | Consequences | Safeguards | Actions |
|---|---|---|---|---|---|
| 1 | More flow | None identified | | | |
| 2 | No flow | Pump P1 stopped | Tank not filled | None | |
| 3 | | Valve V1 shut | Tank not filled | None | |
| | | | P1 damaged, possible fire or explosion | None | Install a pump temperature trip |
| 4 | High level | Delay stopping P1 | Tank overflows. Possible explosion | Level gauge monitored by Operator | Install a high level alarm and high-high level trip that stops P1 |

cover the main scenarios that are known to result in major accidents and significant operational problems. The aim is to combine parameters and guidewords to obtain meaningful deviations, for example no flow, more pressure, less temperature.

Guidewords vary depending on the industry, but are usually defined in sector guidance. There is a school of thought that says the lists are prescriptive and should not be changed or added to. Kletz did not agree [2].

---

### Text Box 1.3

They are based on long experience but, nevertheless, if you find other words are useful, by all means use them, particularly if you are applying HAZOP outside the type of activity for which it was originally designed.

---

#### 1.1.1.4 HAZOP method

Whilst flexible use of guidewords may be acceptable, only studies that follow the defined method should be considered as a HAZOP. The procedure for each node is as follows:

1. Define the node and its design intent;
2. Work through all combinations of parameter and guideword for the node to identify possible deviations (situations where the node would deviate from its design intent);
3. Identify the potential causes of all meaningful deviations identified;
4. Identify potential consequences of each deviation and cause;
5. Identify existing safeguards in place to control the risks of the cause and consequence;
6. Consider the adequacy of safeguards and possible opportunities to reduce risk; and
7. Record the results for future reference.

HAZOP is performed in a meeting or workshop controlled by a team leader. Nodes are selected and displayed to the team so that they know what is being discussed. In the process industry this is usually achieved by marking the nodes on the Piping and Instrument Diagram (P&ID) using coloured marker pens and laying them out on a table, pinning them to the wall, or issuing duplicate copies so that every member of the team can see them. This can also be done electronically so that the marked up drawing can be displayed using a projector (but hand marking is often found to be easier).

The team leader identifies and summarises the node and encourages the team to consider all possible deviations. A scribe is usually appointed to take detailed notes but he/she is also a contributing member of the team.

The details for the HAZOP method may vary in different industries but the basic method should be the same. For an up-to-date description of the HAZOP method for the process industry, refer to the 'HAZOP Guide to Best Practice' published by the Institution of Chemical Engineers [5].

#### 1.1.1.5 Recording the output from HAZOP

The value of a HAZOP is somewhat limited if its results are not recorded in a useful and accessible way. In the time before personal computers, handwritten notes would have been taken and a final record would be made by typing a report. Whilst this remains an option it is now far more common for a scribe to make records using a computer. Generic spreadsheet or word processing software can be used but proprietary software is available that provides templates and handles follow-up reporting and monitoring.

Whether proprietary or generic software is used the output is captured in a table with columns to record deviations, causes, consequences, safeguards, and follow-up actions. This may not be considered as a particularly fascinating read but having a full record of what was discussed and the basis for evaluations and decisions can be essential in the future to confirm a design is appropriate and demonstrate that risks are As Low as Reasonably Practicable (ALARP). Failure to follow the correct HAZOP method or record the results in full will inevitably cause problems in the future.

Although he started carrying out HAZOPs in the days of handwritten notes, Kletz was well aware of the impact computers would have. His list of issues to consider when selecting software, which is summarised in the following table, is still relevant today [2].

---

**Text Box 1.4**

Some of the factors to be considered when choosing a programme for recording the results of HAZOPS:

- Is it simple to use? How much training is required?
- Is it well proven?
- What are the initial and ongoing costs?
- What is the availability and cost of support?
- Are updates available?
- Is it compatible with other programmes?
- What other studies are included?
- Can it be customised? (e.g. can additional columns be added to indicate items which have to be reported to internal or external authorities?)
- Does it include a comprehensive list of prompts?
- How does it monitor actions and changes?
- How are data on failure rates included?
- Can it be linked to accident databases?
- Does it have a spell-check facility?
- Is it possible to carry out a free text search of reports?

---

## 1.1.2 When is HAZOP carried out?

Although very effective and having wide application, HAZOP is not the only safety study available and it is not always the right method to use. Like many things, it is a case of using the right tool to do the right job at the right time.

### 1.1.2.1 Original role of HAZOP on new projects

Kletz was very clear about the timing of a HAZOP [2].

**Text Box 1.5**

A HAZOP cannot be carried out before the line diagrams, complete with control instrumentation (that is, process and instrumentation diagrams) are complete. It should be carried out as soon as possible thereafter, before detailed design starts. The 'window of opportunity' is thus limited, so plan the meetings well in advance. It is no use waiting until the line diagrams are ready and then expecting the members of the team to be available.

If an existing plant is being studied the first step is to bring the line diagrams up to date or check that they are up to date. Carrying out a HAZOP on an incorrect line diagram is the most useless occupation in the world. It is as effective as setting out on a journey with a railway timetable 10 years out of date.

The timing of HAZOP was formalised by ICI (the company where HAZOP originated) with this six stage process:

1. Exploratory phase—identification of basic hazards and review of sites, check for availability of data;
2. Flowsheet phase—coarse HAZOP using Process Flow Diagrams (PFD), identification and assessment of significant hazards;
3. Detailed design—HAZOP on 'frozen' P&IDs;
4. Construction—check that decisions made in earlier studies have been implemented, including hardware and software;
5. Commissioning—precommissioning check and final inspection; and
6. Postcommissioning—safety audit and review after a few months of operation.

The general expectation was that a full HAZOP was performed once.
Kletz highlighted the challenges of including HAZOP in a project plan [2].

**Text Box 1.6**

The HAZOP on a large project may take several months, even with two or three teams working in parallel on different sections of the plant. It is thus necessary to either:

(a) Hold up detailed design and construction until the HAZOP is complete; or.
(b) Allow detailed design and construction to go ahead and risk having to modify the detailed design or even alter the plant when the results of the HAZOP are known.

Ideally, the design should be planned to allow time for (a) but if completion is urgent (b) may have to be accepted.

### 1.1.2.2 Current approaches to integrating HAZOP into new projects

HAZOP may not be defined as a legal requirement anywhere in the world but it has been accepted as a standard method. It is recognised as a valid technique in international standards [6] and it is mentioned very widely in guidance from regulators, institutions, and trade bodies.

In other words, companies are free to decide if a HAZOP is required, but if they decide not to perform one or choose to use a different method they had better have a good explanation about why their approach was better.

All significant projects in the process industry are now usually run in stages, often completed by different teams or contractors. As the project progresses through each stage the quality of information and detail improves. But the same issues remain about doing HAZOP too early or too late. One way of overcoming this is to repeat HAZOP several times during a project. The idea being that information about deviations, causes, consequences, and safeguards can be continually improved as the design develops.

As Kletz said, there is no value in doing HAZOP too early. For example during the earliest phase of a project (concept development) the process and its control are not sufficiently developed. At this time a Hazard Identification (HAZID) (another multidisciplinary workshop-based study) is more appropriate for identifying the major hazards and predicting their possible consequences so that design requirements can be identified. HAZOP is used later in the project to provide the detail of how these identified hazards may be realised and controlled, and to identify any that were not evident at the HAZID.

The first realistic opportunity to perform a worthwhile HAZOP is during Front End Engineering Design (FEED) based on PFDs or preliminary P&IDs. It is performed at a high level using a limited set of guidewords. To be useful a basic control logic should be available.

The most important HAZOP is likely to be performed during detailed design, when P&IDs are mature enough that they can be 'frozen', so that design changes cannot be made until the HAZOP is complete.

HAZOP reports are often a key document in the handover between the teams or contractors performing each phase of the project. They are a valuable reference to show what issues were considered and why the design was considered to be fit for purpose.

### 1.1.2.3 Retrospective HAZOP

There can be a number of reasons why a HAZOP may be carried out retrospectively, for a system that already exists and is in operation. It may be because a HAZOP was not carried out when it was designed, or that changes have occurred which means that the existing HAZOP is no longer representative of its actual operation. Another reason could be to thoroughly check an acquired plant.

A retrospective HAZOP can be carried out in the same way as during design but this may not always be the most effective approach. Having practical experience of operation can help to optimise the process and reduce some uncertainty.

HAZOP can be time consuming and require a lot of resource. Other methods are available that make more use of the additional information available for a retrospective study that may be more appropriate. Process Hazard Review (PHR) is one example [7]. It was also developed by ICI and uses a similar guideword approach. It examines the system at a higher level instead of the detailed nodes used by HAZOP and is reported to be up to five times quicker [7].

### 1.1.2.4 Reviewing previous HAZOP reports

It was common in the past to perform a HAZOP during design, implement any actions, and then file the report away in the archives. This is no longer considered as acceptable and periodic reviews should be carried out to ensure that the study remains relevant and useful. Industries required by law to generate Safety Cases or Safety Reports will usually be expected to review and update these every 4 or 5 years. This seems to be a reasonable frequency for reviewing and updating a HAZOP study and its report.

Things to look for when reviewing a previously completed HAZOP include:

- Creeping changes at the facility due to lots of small hardware and software modifications;
- Changes in the way the system is operated or tasks are performed;
- Evidence from incident reports and other sources that may suggest the HAZOP overlooked certain risks or that the specified controls are inadequate; and
- New information becoming available about hazards and the development of new good practices in industry.

If only minor issues are uncovered during a review it is normally acceptable to update the report, highlighting any changes using standard document control processes. If major issues are found it may be necessary to conduct a retrospective HAZOP or to carry out an alternative assessment for the same purpose such as PHR [7].

### 1.1.3 Why HAZOP?

Kletz explained the reason why HAZOP had to be developed [2].

---

**Text Box 1.7**

The traditional method of identifying hazards—in use from the dawn of technology until the present day—was to build the plant and see what happens—'every dog is allowed one bite'. Until it bites someone, we can say that we did not know it would. This is not a bad method when the size of an incident is limited but is no longer satisfactory now that we keep dogs which may be as big as Bhopal (over 2000 killed in one bite) or even Flixborough (28 killed). We need to identify hazards before the accidents occur. Check-lists are often used to identify hazards but their disadvantage is that items not on the list are not brought forward for consideration and our minds are closed to them. Check-lists may he satisfactory if there is little or no innovation and all the hazards have been met before, but are least satisfactory when the design is new.

---

These days, plants are not built 'to see what happens' and there are very clear duties on companies to do everything reasonably practicable to identify hazards and control risks. HAZOP is probably the most recognised method available to assist with this and there is a high expectation that a HAZOP is carried out for any new major hazard facility, unless another equally effective method is used.

### 1.1.4 Alternatives

Motivational theorist Abraham Maslow is quoted as saying "If the only tool you have is a hammer, you will see every problem as a nail." HAZOP is one of our better tools but we have many others available. These include [8]:

- Previous incident reports and analyses;
- Human factors and error assessment;

- Integrity management and review (particularly for aging facilities);
- Reliability analysis;
- Dynamic process simulation;
- Inherent safety review;
- ALARP reviews;
- Process materials and fluids analysis;
- Barrier analysis; and
- Safety critical element reviews and performance standards.

A study of hazard identification methods [9] identified 40 but concluded that HAZOP was one of only a few with any formal guidance. Also, it can be seen that some of the other methods available are adaptations of the standard HAZOP.

### 1.1.5 Accidents that should have been prevented by HAZOP

Kletz dedicated a whole chapter in his book about HAZOP [2] to describe "some accidents that could have been prevented by HAZOPs." At the time he wrote this it was probably fair to say that HAZOP was relatively uncommon. This meant that Kletz could speculate that an accident happened because of a lack of risk controls because a HAZOP had not been carried out. Things have changed and now the expectation is that HAZOP is carried out and so for more recent accidents it is more reasonable to consider why the HAZOP (or equivalent method) was not effective.

#### 1.1.5.1 Accidents due to 'reverse flow'

Kletz identified [2] that

---

**Text Box 1.8**

Many accidents have occurred because process materials flowed in the opposite direction to that expected and the fact that this could occur was not foreseen. For example, ethylene oxide and ammonia were reacted to make ethanolamine. Some ammonia flowed from the reactor, in the wrong direction, along the ethylene oxide transfer line into the ethylene oxide tank, past several nonreturn valves and a positive pump. It got past the pump through the relief valve which discharged into the pump suction line. The ammonia reacted with $30\,m^3$ of ethylene oxide in the tank which ruptured violently. The released ethylene oxide vapour exploded causing damage and destruction over a wide area.

A hazard and operability study would have disclosed the fact that reverse flow could occur.

On another occasion some paraffin passed from a reactor up a chlorine transfer line and reacted with liquid chlorine in a catchpot. Bits of the catchpot were found 30 m away.

On many occasions process materials have entered service lines, either because the service pressure was lower than usual or the process pressure was higher than usual. The contamination has then spread via the service lines (steam, air, nitrogen. water) to other parts of the plant. On one occasion ethylene entered a steam main through a leaking heat exchanger. Another branch of the steam main supplied a space heater in the basement of the control room and the condensate was discharged to an open drain inside the building. Ethylene accumulated in the basement, and was ignited (probably by the electric equipment, which was not protected), destroying the building. Again, a HAZOP would have disclosed the route taken by the ethylene.

---

### 1.1.5.2 Texaco Pembroke Refinery (1994)

On 24 July 1994 a severe electrical storm caused major plant disturbances across the whole refinery [10]. Whilst attempting to resume normal operations a large quantity of hydrocarbon was routed to the flare system. Liquids collected in the flare knock-out drum and eventually overflowed into the downstream flare pipework, which failed leading to a significant release. This ignited and exploded, leading to significant damage to the site and beyond. There were some injuries but fortunately no one was killed.

The flare knock-out drum had originally been designed so that any liquid collected would drain away automatically to slops. Modifications had been made to return the liquid to the process in order to reduce waste, but which significantly reduced the rate that liquid could be removed from the drum. A HAZOP of this modification should have identified that during major plant upsets the flow of liquid to the drum would have been greater than the capacity of the new draining system and also, that high level in the drum could result in liquid break-through into the downstream pipework, which was not designed to handle this (Fig. 1.2).

### 1.1.5.3 Electrostatic precipitator explosion

An explosion occurred on 18 February 2015 at a refinery in Torrance, California [11]. A pressure deviation during preparation for maintenance allowed hydrocarbon to backflow into an electrostatic precipitator where it ignited and exploded, damaging the equipment and injuring four people in the area. The hydrocarbon was able to backflow because a slide valve, which was being used as the isolation, had been eroded by solid catalyst that passed through

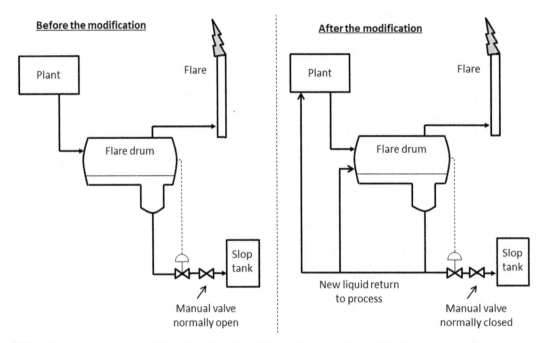

FIG. 1.2    The arrangement of flare drum liquid handling before and after modifications.

it during normal operation. This valve had been identified as safety critical but the steps taken by the refinery to prove its integrity only confirmed that mechanical components were working and not whether it was effective as an isolation.

A HAZOP of this system should have identified that pressure deviations could occur and that these would result in hydrocarbon back-flowing through the slide valve to the electrostatic precipitator where ignition was likely to have occurred. It should have highlighted that an effective isolation would be required to avoid this and that a single slide valve that was used for control was very unlikely to be sufficient for this purpose.

### 1.1.5.4 *Esso Longford*

Kletz referred to the Esso Longford accident to illustrate how HAZOP could have prevented it [12].

---

**Text Box 1.9**

On 25 September 1998 a heat exchanger in the Esso gas plant in Longford, Victoria, Australia fractured, releasing hydrocarbon vapours and liquids. Explosions and a fire followed, killing two employees and injuring eight. Supplies of natural gas were interrupted throughout the State of Victoria and were not fully restored until 14 October. There was no alternative supply of gas and many industrial and domestic users were without fuel for all or part of the time that the plant was shut down.

Many companies have examined their older plants to judge their fitness for purpose, including the suitability of the materials of construction. Some equipment is replaced so that it can withstand fault conditions. In other cases control and trip systems are improved, instructions revised or the need to avoid low temperatures emphasised in operator training. Esso, part of the US Exxon company, decided to carry out hazard and operability studies (HAZOPs) on all the old plants on their Longford site. Most of them were completed but the 1969 plant, the oldest on the site, was never studied. According to the official Report (p. 203), 'no satisfactory reason was given in evidence for its deferral or abandonment'. A HAZOP would almost certainly have disclosed that the heat exchanger that failed could become too cold and would have led to a systematic examination of the ways in which this could occur and the actions that might be taken to prevent it. It would also have disclosed the need to provide better isolations in the connecting lines to the other units.

---

## 1.1.6 Ensuring effective HAZOP

Kletz had some concerns that people underestimated the HAZOP method, viewing it as one of the "hammers and spanners of the chemical industry" [2].

---

**Text Box 1.10**

This makes them sound too easy. They are, in fact, sophisticated tools which require long, detailed application. They are not quick fixes.

---

He saw that part of the problem was that, as it has become ubiquitous, people would assume that studies would automatically become 'better and better'. His response to this view was as follows [2].

---

**Text Box 1.11**

In some ways this is true; the best are getting better. However, as with any technique, as more and more people start to use it the standard can easily fall and it can become a perfunctory exercise, carried out because the company rules (or the regulators) require it and not because those present really feel that it is worthwhile. Some auditing of the quality of HAZOPs is therefore desirable.

---

The key message is that HAZOP requires respect and resources. Companies should have procedures in place that define the requirements, use competent team leaders, and always involve the right people. As with all processes, continual review and audit is required. Kletz gave this perspective [2].

---

**Text Box 1.12**

As the use of any technique becomes more widespread its quality is liable to decrease. There is therefore a need to be able to audit the quality of a HAZOP. At a workshop on HAZOP held in 1995 the auditing of HAZOP was selected as the most pressing current topic. The best method of auditing is to sit in on a HAZOP. If that is not practicable—for example, because the HAZOP is complete—Rushton[b] has described an audit scheme. The auditor samples the documentation produced by the HAZOP and looks for evidence that various modes of operation—such as start-up, shutdown and maintenance—have been considered in addition to normal operation, that the knowledge and experience of the team members were adequate, that the same people attended throughout and did so regularly, that the recommendations made were carried out, and that any late changes in design were studied. If the plant has already been commissioned the auditor should examine the problems that have arisen and see if they could reasonably have been spotted during the HAZOP. Altogether there are six pages of suggested questions. The auditor should talk to the team members to gain their impressions and assess their knowledge and experience.

---

## 1.2 HAZOP organisation and process

Although HAZOP provides a method for performing a hazard analysis it relies on people to identify the relevant deviations, recognise potential hazards, and make reasonable judgements about existing safeguards and potential new ones.

---

[b] A. Rushton, Quality assurance of HAZOP. Health and Safety Executive offshore technology report OTO 96 002. 1996—(Ref. [13] in this compendium).

## 1.2.1 Organising a HAZOP

From a technical perspective a HAZOP cannot take place until there is enough information about the system being assessed. For process plant this is when an up-to-date and 'frozen' set of P&IDs are available. Once this is achieved there are three key factors for a successful HAZOP:

- A committed, knowledgeable, and creative team;
- An experienced and competent Team Leader; and
- Meeting organisation.

### 1.2.1.1 HAZOP team members

The make-up of the team has a significant impact on the quality of a HAZOP. The challenge is to get a group of people together who cover a range of knowledge and experience whilst keeping the process manageable. Also, to involve people who can be creative and are willing to discuss the issues.

Kletz identified the following people to be involved in a HAZOP for a new facility [2].

---

**Text Box 1.13**

**Project or design engineer**

Usually a mechanical engineer and, at this stage of the project, the person responsible for keeping the costs within the sum sanctioned. The project engineer wants to minimise changes but at the same time find out now rather than later if there are any unknown hazards or operating problems.

**Process engineer**

Usually the chemical engineer who drew up the flowsheet.

**Commissioning manager**

Usually a chemical engineer, the commissioning manager will have to start up and operate the plant and is therefore inclined to press for any changes that will make life easier.

**Control system design engineer**

Modern plants contain sophisticated control and trip systems and HAZOPs often result in the addition of yet more instrumentation.

**Research Chemist**

If new chemistry is involved.

---

The importance of the HAZOP in verifying the safety and operability of the plant and the expense of modifications means that the Project Manager will usually attend for some if not all of the HAZOP.

Whilst Kletz suggested that 'modern plants' would require a higher input from control systems engineers this may no longer be the case because most control software is usually not bespoke. However, knowledge of the control logic is required including any role it has in safeguarding to prevent or mitigate hazardous scenarios.

Kletz pointed out that the make-up of the team may vary according to the nature of the design process, for example [2].

**Text Box 1.14**

If the plant has been designed by a contractor, the HAZOP team should contain people from both the contractor and client organisation, and certain functions may have to be duplicated.

HAZOP practices have evolved and the team make-up has probably changed since Kletz first started writing on the subject. For example, it is increasingly recognised that involving experienced operators and maintenance technicians is a critical factor in achieving a good HAZOP. They are the people who know what actually happens, what works, and what doesn't. This is particularly the case for a retrospective or modification HAZOP, but even for a completely new facility practical experience can provide a real insight into potential issues that designers and project engineers might not recognise. Also, people who have spent their working lives at the sharp end of operations have usually developed robust and effective communication skills that can help the whole team focus on the critical issues and avoid getting hung up on trivia and interesting but irrelevant technical details.

Another key member of the HAZOP team, who was not mentioned by Kletz, is the Scribe. They are usually a junior process engineer. Their role is to record the HAZOP inputs and outputs, but they are still a full team member and should be able to contribute to the technical discussions. They need to understand the discussions that take place, be reasonably quick at taking notes (typically directly on a computer nowadays), and be prepared to slow down the meeting and ask for clarification if required [14]. The scribe usually does most of the organisational legwork, for example assembling documentation. It has been reported [15] that "a good scribe can cut 30% of the time taken to complete a HAZOP, whereas a bad scribe can virtually wreck it."

One of the concerns of inviting people who are not usually directly involved in design (particularly operators and maintenance technicians) is that they are not familiar with the method. In fact some people invited to a HAZOP may be worried that they may struggle to understand what is going on. Kletz pointed out that team members are invited due to their knowledge and experience and are not expected to be experts at HAZOP [2].

**Text Box 1.15**

HAZOP teams, apart from the team leader, do not require much training. They can pick up the techniques as they go along. If anyone is present for the first time, the team leader should start with 10 min of explanation. However, if possible, new team members should attend a half-day lecture and discussion based on this chapter. The Institution of Chemical Engineers can supply a training package. The team leader should, however, start the discussion of each line or plant item by explaining, or asking someone to explain, its purpose.

Another key point, highlighted by Kletz, is that the value of the input made by team members is not restricted to their technical knowledge. In fact it is often their willingness to discuss the things that they do not understand that can add the most value. It is worth noting that the designers should have already considered the well-known and understood hazards and operability issues and so it is the quirky and unusual potential issues that are of most concern [2].

## Text Box 1.16

It might be thought that membership of a HAZOP team is "the proper toil of artless industry, a task that requires neither the light of learning, nor the activity of genius, but may he successfully performed without any higher quality than that of bearing burthens with dull patience and ... sluggish resolution," to quote Dr. Johnson.[c] This is not the case. The best team members are creative and uninhibited people who can think of new and original ways for things to go wrong and are not too shy to suggest them. In a HAZOP, do not hesitate to suggest impossibly crazy deviations, causes, consequences or solutions as they may lead other people to think of similar but possible deviations, etc. Zetlin writes. "I look at everything and try to imagine disaster. I am always scared. Imagination and fear are among the best engineering tools for preventing tragedy."[d]

Another feature of good team members is a mental ragbag of bits and pieces of knowledge that they have built up over the years. Such people may be able to recall that a situation similar to that under discussion caused an incident elsewhere. They need not remember the details so long as they can alert the team to possibilities that should be considered and perhaps investigated further.

A final point made by Kletz is that the status of the team and its members can have a big impact on a HAZOP [2].

## Text Box 1.17

The team should have the authority to agree most changes there and then. Progress is slow if every change has to he referred to someone who is not present. The team members should try to avoid sending deputies. They lack the knowledge of previous meetings and might not have the authority to approve changes: as a result progress is held up. Some people have told me that this is impracticable in their companies as all changes have to be approved at a high level. This does not matter so long as the team members feel confident that most of their recommendations will be accepted without argument. However, if the discussions in the HAZOP meetings have to be gone through again, time is wasted.

### 1.2.1.2 Team leader

The Team Leader is critical for any HAZOP and their ability to use the method, control the workshop, and manage the overall process will have a significant impact on the quality of the outcomes. Kletz summed up the main requirements [2].

## Text Box 1.18

**Independent team leader**
An expert in the HAZOP technique, not the plant. The job of the team leader is to ensure that the team follows the procedure. To be a successful team leader you need to be skilled in leading a team

[c]S. Johnson. A dictionary of English Language. 1755.

[d]L. Zetlin. Quoted by H. Petrowski in Design Paradigms. Cambridge University Press. 1994.

of people who are not responsible to you, and be the sort of person who pays meticulous attention to detail. It is easy to underestimate the ability required. It is not a job that anyone can do. The team leader may also supply the safety department's view on the points dictated. If not, a representative from this department should be present.

The Team Leader needs to ensure that the HAZOP participants play their part in the workshop. They need to facilitate the discussion so that everyone is actively involved and is prepared to discuss issues without becoming side-tracked. All contributions should be constructive so that solutions are identified and not just potential problems.

The Team Leader should be independent of the project and some companies insist that they are provided by an external organisation (i.e. a consultant). The Team Leader should have received some formal, certified training.

Other traits of a good Team Leader include knowledge of other methods (e.g. Failure Modes and Effects Analysis (FMEA), Layer of Protection Analysis (LOPA), HAZID, attention to detail, and enough stamina to adhere to the method for the full duration [14]. They need to be able to handle people, particularly the dominant personalities that may be inclined to take over the meeting and dissuade others from talking. This can be a particular problem if this person is an engineer who is overly defensive of their design.

Kletz pointed out that motivating the team is vitally important [2]:

---

### Text Box 1.19

Although HAZOP is a valuable technique, no-one jumps out of bed on a Monday morning shouting, 'Hooray! I've got a HAZOP today!'. The need to consider every deviation on every line can become tedious. Beware of making it more so by bureaucratic procedures such as insisting on excessive recording or discussing everything twice (or three times)—in the HAZOP meeting and afterwards with the boss of the project team. There is a net loss if, in our eagerness to document everything to explain it to everybody, we discover less information worth documenting. If HAZOP and similar systems are not acceptable to creative minds, they will never succeed.

---

It is possible for the Team Leader's involvement to be limited to the duration of the workshop, with a short time later to review a report prepared by the Scribe or others. In this case they are relying on others to prepare and organise the HAZOP. This may be the case if an external consultant has been engaged for the role. In practice the HAZOP will generally run more smoothly and efficiently if the Team Leader has oversight of the whole organisation and is actively involved in planning the time requirements, selecting the team, establishing the support requirements, and developing the terms of reference for the study [16]. This is important because resources are always limited and any inefficiency will reduce the time available for conducting the HAZOP and will inevitably effect the scope covered and/or depth of assessment carried out.

### 1.2.1.3 *Meeting organisation*

HAZOP is performed by a group of people working together in a meeting or workshop. The way this meeting is conducted will have a significant impact on the quality of the assessment and output.

The factors that affect a HAZOP meeting are largely the same as any other type of meeting. A shared understanding of objectives amongst attendees, good preparation and time management, strong leadership, and all participants taking an active and pragmatic approach will have positive influences. A suitable venue and equipment will help the meeting to progress and management commitment is essential to ensure attendance of the right people at the right time [14].

Kletz highlighted that adhering to a systematic approach is vital [2]. The meeting must be managed to avoid aimless discussion and straying off on tangents.

---

**Text Box 1.20**

The complexity of modern plants makes it difficult or impossible to see what might go wrong unless we go through the design systematically. Few accidents occur because the design team members lack knowledge; most errors in design occur because they fail to apply their knowledge. HAZOP gives them an opportunity to go through the design line by line, deviation by deviation, to see what they have missed.

---

The time required to conduct a HAZOP is often a bone of contention. Kletz tended to take the view that it will take as long as it requires and advocated that the meetings themselves should be limited in their duration and frequency to avoid fatigue of team members and the temptation to rush the HAZOP [2].

---

**Text Box 1.21**

A HAZOP usually takes 1.5–3 h per main plant item (still, furnace, reactor, heater, and so on). If the plant is similar to an existing one it will take 1.5 h per item but if the process is new it may take 3 h per item. Inexperienced teams of course, take longer than experienced ones. Meetings are usually restricted to 3 h, 2 or 3 days per week to give the team time to attend to their other duties and because the imagination tires after 3 h at a stretch. If the members of the team have to be gathered from a distance, longer periods of working, perhaps every morning for a week, may have to be accepted. Resist any temptation to work 8 or more hours per day for a week, as attention inevitably flags. It is the results of a HAZOP that are important, not the number of hours spent on it.

---

The times proposed by Kletz may have been a bit optimistic because a complicated system such as a reactor might take a lot longer than 3 h to complete. Using an average of 4–5 h per P&ID or an average of 3 h per subsystem may provide a reasonable estimate for planning. However, planning should include input from an experienced team leader who will be able to give more accurate estimates of time, which can avoid arbitrary plans being set based on resources perceived to be available [16].

The suggestion that HAZOPs should not occupy full, consecutive days remains appropriate and valid. Unfortunately it is a guideline that is often overlooked, particularly in major projects. The problem is that adhering to these restrictions will have a significant impact on schedule. If the HAZOP results in a delay in project delivery it will be very unpopular and the impact that this has on perceptions of the role of process safety and reasonableness of approach may be counter-productive. If the schedule dictates that full and consecutive days are required, then the potential issues with fatigue of team members and hence the quality of the HAZOP output need to be factored into the overall approach being taken to address process safety.

The approach taken by the Team Leader can have a significant effect on how the HAZOP meeting progresses. Kletz identified a range of issues that they need to manage [2]:

---

**Text Box 1.22**

While the team members have a common objective—a safe and operable plant—the constraints on them are different. The designers, especially the design engineer responsible for costs, want to keep the costs down. The commissioning manager wants an easy start-up. This conflict of interests ensures that the pros and cons of each proposal are thoroughly explored before an agreed decision is reached. However, if the design engineer has a much stronger personality than the other members, the team may stray too far towards economy. Other teams may err the other way. The team leader tries to correct any imbalance. To quote Sir John Harvey-Jones,[e] "In industry the optimal level of conflict is not zero."

---

One of the most common causes of disruption of the team and inefficient HAZOP is spending too much time on discussion of design changes. The HAZOP is not a design meeting. If the solution to an identified problem is not obvious, the Team Leader must quickly guide the team to formulate an action and move on [2].

---

**Text Box 1.23**

If the team cannot agree, the team leader should suggest that the point is considered outside the meeting. Sometimes a decision is postponed while expert advice is sought—for example, from a materials expert—or even while research is carried out. Sometimes a decision is postponed so that a quantitative estimate of the hazard can be made.

---

One common failure from HAZOP (and other safety studies) is poorly defined actions. For example:

- Actions that ask people to 'Consider' an issue without any clear acceptance criteria so that they can be closed without any action taken or any justification;
- Actions that make sense when raised during the HAZOP but on their own are difficult to interpret or are ambiguous outside of the meeting; and
- Actions that are too specific so that they can be closed by a simple design change that does not address more fundamental and wide-ranging issues.

Actions from HAZOPs must be "written sufficiently clearly that the discipline engineers implement the intent of the action; and not just the title" [15]. Sometimes it is not possible to formulate the wording correctly in the workshop and it is usually acceptable for the Team

---

[e]Sir John Harvey Jones was a flamboyant Chairman of ICI and broadcaster.

Leader and Scribe to have a wash-up session afterwards to finalise details including wording of actions. This should be done very soon after the workshop with any changes communicated to the HAZOP team to allow them to comment.

Overall, it must be recognised that HAZOPs take up significant resources. The people who attend will usually have lots of other work to be done, whether that be progressing the design for a new project or operating an existing facility. Whilst the HAZOP is important, there can be no justification for using time inefficiently due to poor organisation and planning. Identifying in advance similar nodes or areas where the HAZOP will become very repetitive can allow for a different approach to be taken to maintain interest and attention of the team so that they will continue to pay close attention and be actively involved.

## 1.2.2 Accidents that *may* have been avoided by HAZOP

Although HAZOP has been proven to be effective, nevertheless it is reliant on the people involved. Whilst guidewords can prompt them to think about a range of scenarios, if potential hazards are not understood it is possible that some will not be recognised. Having the right people involved in the HAZOP certainly reduces the risks but this cannot be guaranteed.

### 1.2.2.1 Abbeystead

Although included in his chapter about accidents that *could* have been prevented by HAZOP, Kletz highlights that at Abbeystead this would only have been the case if someone in the team had the insight to recognise that methane could be present [2].

---

**Text Box 1.24**

At Abbeystead, water was pumped from one river to another through a tunnel. In an incident in May 1984, when pumping stopped, some water was allowed to drain out of the tunnel leaving a void. Methane from the rocks below accumulated in the void and, when pumping was restarted, was pushed through vent valves into a pump house where it exploded, killing 16 people, most of them local residents who were visiting the plant.

If anyone had realised that methane might he present, the explosion could have been prevented by keeping the tunnel full of water or by discharging the vent valves into the open air. In addition, smoking, the probable cause of ignition, could have been prohibited (though we should not rely on this alone). None of these things were done because no-one realised that methane might be present. Published papers contain references to the presence of dissolved methane in water supplies but these references were not known to the water supply engineers. The knowledge was in the wrong place.

Could a HAZOP have prevented the accident? Only if one of the team knew or suspected that methane might be present. He need not have known the details so long as he could recall the fact from the depths of his memory. Good HAZOP team members are people who have accumulated, by experience and reading, a mental ragbag of bits and pieces of knowledge that may come in useful one day. A HAZOP provides opportunities for the recall of long-forgotten bits of knowledge that might otherwise never pass through the conscious mind again.

---

### 1.2.2.2 Anhydrous ammonia leak

On 23 August 2010 a release of anhydrous ammonia at the Millard Refrigerated Services facility in Theodore, Alabama resulted in 32 people being admitted to hospital, with four of those requiring intensive care [17]. It happened because pipework failed due to hydraulic shock during the restart of the plant's ammonia refrigeration system following a 7-h power outage. One of the causes was identified as "the control system contained a programming error that permitted the system to go from soft gas directly to refrigeration mode without bleeding the high pressure from the coil or preventing the low-temperature suction valve from opening. The error with the software logic in the control system went undetected because under normal operations, in its programmed sequence, the defrost cycle would not allow the ammonia liquid to enter the evaporator until the coil was properly depressurized via the bleed cycle."

A HAZOP should have identified that an interruption in the plant's automatic defrost cycle could result in steps occurring out of sequence, some of which would be hazardous. This alone may not have been sufficient to prevent this accident because the cause was a software error. However, the HAZOP may have prevented the accident if the criticality of the software controlled cycle had been fully understood so that it was subject to more scrutiny during its initial programming and any subsequent modification. It is very common, especially for legacy systems, for critical steps to be hidden within software code that is predominantly concerned with normal operations. As a minimum, a functional specification should define all safety critical elements so that they can be actively confirmed as part of initial acceptance tests and during management of change.

### 1.2.2.3 Heat exchanger rupture caused by High Temperature Hydrogen Attack

In April 2010 a heat exchanger ruptured at Tesoro's Anacortes refinery in Washington State, United States resulting in a fire and explosion. Seven people died [18]. The rupture occurred due to a reduction in the strength of steel caused by High Temperature Hydrogen Attack (HTHA). This occurs at high temperatures when hydrogen is present.

HTHA is a known issue and it is reasonable to expect a HAZOP team to have some awareness. But it only occurs under certain conditions. In this case it was concluded that the exchanger should have operated outside of the region where HTHA was a significant risk, but a HAZOP team member with a more in-depth knowledge may have raised the issue and insisted that the exchanger was subject to postweld heat treatment, which can reduce susceptibility. Also, they may have insisted that active monitoring of the exchanger's operating condition was to be considered critical in order to detect any deviations from the predicted range into a region where HTHA was a risk.

## 1.2.3 Harnessing available knowledge

The accidents described previously highlight that HAZOP is most effective if knowledge from a range of sources can be harnessed effectively. Kletz highlighted a number of different examples [2]. The first case is a floating roof tank located in a bund, shown in Fig. 1.3.

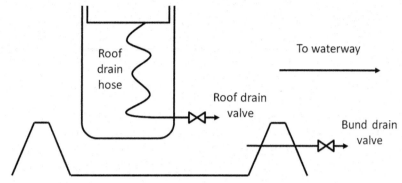

FIG. 1.3    Considering the potential for a hose to leak.

**Text Box 1.25**

The tank contains oil. Rainwater can be drained from the roof into the bund and from there into the main drainage system. Suppose a HAZOP team is considering whether any substance 'other than' water can get into the main drainage system. For this to occur there would have to be a hole in the hose, and both valves would have to be left open. An inexperienced team might decide that a triple coincidence is so improbable that there is no need to consider it further. However, someone with knowledge of the practicalities of plant operation would realise that during prolonged rain the operators may leave both drain valves open, whatever the instructions say, to avoid frequent visits to the tank. Any hole in the hose will then contaminate the main drainage system with oil.

Kletz used the following example to emphasise the need for specialised knowledge [2].

**Text Box 1.26**

A vessel contained liquid sulphur (melting point 120°C). A HAZOP was carried out on the flowsheet; the team considered 'more of pressure' and decided that the precautions taken to prevent choking of the vent, which included a lute, were adequate. At a later HAZOP of the line diagram, when considering 'more of temperature', someone pointed out that the viscosity of sulphur rises sharply above about 200°C. This temperature could not be reached in normal operation but could be reached if the vessel was exposed to fire. The sulphur in the lute could then become so viscous that it would prevent relief of the vessel. The relief system had to be redesigned.

A solvent tank was vented through a seal pot. An electric heater was added later. The reason is not stated in the report, but was presumably to prevent freezing in cold weather. The modification was HAZOPed but all the members of the team were chemical engineers; no electrical engineer or representative of the supplier was present. None of the chemical engineers realised that the temperature of the heater could rise above the auto-ignition temperature of the solvent if the liquid level in the seal pot was lost.

Kletz highlighted the importance of local knowledge [2].

---

**Text Box 1.27**

During the HAZOP of a batch process the team asked what might be added to the reactor 'other than' the materials that should be present. The word they actually used was 'contamination'. Someone pointed out that organic acids could cause a runaway. Further discussion revealed the fact that organic acids were used in another process and were stored in the same warehouse and in the same colour and type of drum as one of the reactants.

---

Kletz showed how wide-ranging knowledge of what happens including what other people do can make a big difference in a HAZOP [2].

---

**Text Box 1.28**

A plant was fitted with blowdown valves which were operated by high-pressure gas. On a cold day, a leak on the plant caught fire. The operators isolated the feed and tried to blow off the pressure in the plant. The blowdown valves failed to open as there was some water in the impulse lines and it had frozen. As a result the fire continued for longer and caused more damage than it would otherwise have done.

How the water got into the impulse lines was at first a mystery. At a HAZOP 2 years earlier, when the plant was modified, the team were asked if water could get into the impulse lines and they said 'No'.

Occasionally the valves had to be operated during a shutdown, when no high-pressure gas was available. The maintenance team members were asked to operate the valves but not told how to do so. They used water and a hydraulic pump. None of the HAZOP team members, which included the operator shop steward, knew that the valves had been operated in this way.

HAZOPS are only as good as the knowledge and experience of the people present. If they do not know what goes on, the HAZOP cannot bring out the hazards.

---

## 1.3 Limitations of HAZOP

Although considered to be one of the best tools available, Kletz pointed out that HAZOP has its limitations [2].

---

**Text Box 1.29**

HAZOP cannot, of course detect every weakness in design. In particular, it cannot draw attention to weaknesses in layout. It will also miss hazards due to leaks on lines that pass through or close to a unit but carry a material that is not used on that unit. This can be overcome by using an additional guide word such as PASSING THROUGH or NEARBY.

HAZOP assumes that the design assumptions are followed during construction and operation. If, say, the wrong material of construction is used or equipment is not tested as assumed, then problems may result. HAZOP teams may, of course, draw attention to circumstances where special measures should be taken to ensure that the right materials are used or tests carried out, and may question the wisdom of including equipment such as bypasses around trip valves or isolation valves below relief valves.

Fundamentally HAZOP is limited because it analyses system drawings (e.g. P&ID), which are only a representation of some aspects of the system [15].

## 1.3.1 Poorly executed HAZOP

### 1.3.1.1 *HAZOP as a design review*

Kletz identified a common problem where designers assume problems they encounter can be addressed by a HAZOP [2].

---

**Text Box 1.30**

Design engineers have been known to say, when someone suggests a change in design, "Don't bother me now. We'll be having a HAZOP later on. Let's talk about it then."

---

He points out that this is not the purpose of HAZOP and can result in the assessment being compromised because it is side-tracked into design activities, which means hazards and operability problems can be overlooked [2].

---

**Text Box 1.31**

This is the wrong approach. A HAZOP is a final check on a basically sound design to make sure that no unforeseen effects have been overlooked. It should not replace the normal consultations and discussions that take place while a design is being developed. A HAZOP meeting is not the right place for redesigning the plant; there are too many people present and it distracts from the main purpose of the meeting which is the critical examination of the design on the table.

---

### 1.3.1.2 *HAZOP completed too late*

A fairly common problem is that HAZOP is done too late in a project. This means that people are reluctant to get too involved because they know that if they do identify any problems with the design they will be very difficult and expensive to rectify. This arises when the project is not well managed or the value of HAZOP is not recognised. The study happens because it is a key deliverable that has to be done but gets left to the last minute.

The problem with a late HAZOP is that recommendations for improvements will be mainly restricted to add-on safety protection systems rather than implementing inherently safer solutions. The result is that the hazards remain but the plant complexity is potentially increased due to all the devices being added [19].

### 1.3.1.3 *Short cutting the process*

One of the significant challenges for HAZOP is the resources required. It is inevitable that people will try to modify their approach to reduce this burden.

Kletz identified one example where this was done [2].

---

**Text Box 1.32**

One company has tried to simplify HAZOP by just looking for deviations from its design codes and standards. This may be OK if there is little or no innovation but if there is innovation—and there usually is some—the existing codes may not cover the new circumstances. For example, a hydraulic crane tried to lift a load that was too heavy for the fully extended jib and fell onto the plant: no alarm sounded. The crane was fitted with all the alarms required by the current codes and they were all in working order. However, the codes were written for mechanical strut cranes. Hydraulic cranes have an extra degree of freedom—the length of the jib can be changed—and therefore an extra alarm is needed, but no-one had realised this.

---

One way of short cutting the process is to minimise the amount of analysis conducted in the meeting or workshop. People justify this by saying that their designers are highly competent in both design and HAZOP, and so will have already completed a thorough analysis themselves as they went along. Taken to an extreme this philosophy can lead to a view that HAZOP is not required at all. This is something that Kletz discussed [2].

---

**Text Box 1.33**

'We don't need a HAZOP. We employ good people and rely on their knowledge and experience'.

A HAZOP is no substitute for knowledge and experience. It is not a sausage machine which consumes line diagrams and produces lists of modifications. It merely harnesses the knowledge and experience of the team in a systematic and concerted way. Because designs are so complicated the team members cannot apply their knowledge and experience without this crutch for their thinking. If the team lacks knowledge and experience the HAZOP will produce nothing worthwhile.

'Good People' sometimes work in isolation. Pegram writes, "working independently, the solving of a problem by one discipline can become a problem of another" and "low cost engineering solutions from one point of view may not necessarily end up as overall low cost." HAZOP ensures that hazards and operating problems are considered systematically by people from different functions working together. Experience shows that start-up, shutdown and other abnormal conditions are often overlooked by functional groups working in isolation.

The opposite of the heading to this section is the belief that good systems can be a substitute for good people. All that systems can do, however, is to ensure that people's knowledge and experience are applied systematically and thus reduce the chance that something is missed. If people lack knowledge or experience (or commitment) then systems such as HAZOP are empty shells. People will go through the motions but the output will be poor. Good people without a system will achieve less than their full potential, but if people lack knowledge and experience the systems will achieve nothing.

---

In fact, one flip side of involving knowledgeable and experienced people in HAZOP is that they can provide convincing arguments to say an issue is not a major concern because they have seen it happen and the consequences were not significant [15]. In this case the experience is helping to keep the HAZOP moving along at a good pace but introduces the potential for things to be missed, particularly the low-frequency, high-consequence events. Another way that this can happen is if the team get fixated on one or two major hazards, because then they can start to perceive that lesser ones are not worth discussing. But as we know most accidents occur due to a combination of multiple, relatively minor issues and not just one very large one in isolation.

## 1.3.2  Inappropriate safeguards

It is tempting in a HAZOP to identify a range of existing items as Safeguards or to propose additional items without taking account of what they can reasonably achieve in practice. For example [20]:

- Local instruments which are never checked by field operators and, therefore, could in no way be considered safeguards;
- Alarms which give the operator insufficient time to effectively halt the deviation, because the rate of upset is too fast. Also, alarms that are too generic, activated frequently, or would occur as part of a flood of alarms in the scenario being considered;
- Pressure relief systems for which there is no guarantee that they were designed for the case being studied; and
- Operating procedures, when the cause giving rise to the scenario is human error.

### 1.3.2.1  *Over reliance on people*

As with all things in life it is necessary to take a balanced approach to HAZOP, particularly when deciding what safeguards are required and which are not because they would not add additional benefit. When making these decisions it is important to recognise the ability of the people involved in operating the system, but to be realistic about their capabilities and reliability. This can be particularly difficult if experienced operators are involved in the process, if they become protective of their role, as they are likely to find it difficult to believe that they or their colleagues will make the errors that would result in the deviations being discussed. This is often a wider cultural issue caused by operators being unfairly blamed for incidents and other problems experienced.

Kletz pointed out that human errors must be considered fully [2].

---

Text Box 1.34

Human error should also always be considered as a possible cause of deviations. Thus no flow may be due to someone failing to open a valve. This can occur for a number of reasons.

The operator may not have known the valve should have been opened; the intention was wrong. We may have to improve training and instructions or simplify the job.

The operator may have decided that it was unnecessary to open the valve at that time or that other tasks were more urgent. We may have to explain the reasons for instructions and make sure they are followed.

The valve may have been too stiff or out of reach.

Most likely of all, there may have been a slip or lapse of attention, the intention was correct but was not fulfilled. Everyone has slips and lapses of attention from time to time and they cannot be prevented, though various actions may make them less likely. If the consequences are serious, we should remove or reduce the opportunities for error by changing the design or method of working (or protect people from the consequences or make recovery possible).

---

There are two fairly common problems that result from a lack of understanding of human factors when carrying out a HAZOP:

- Discounting deviations because it is assumed that competent people who follow procedures will not make errors; and
- Identifying procedures as effective safeguards without specifying exactly what procedure is required and what content is required to ensure that the safeguard is effective.

Although it is possible to take some credit for procedures in a HAZOP the focus must remain on finding design solutions. The HAZOP team have no control over what procedures are provided when the system is operational or the competence of the people who operate, maintain, and manage it.

### 1.3.2.2 *Unrealistic expectations of automated systems*

A very common outcome from a HAZOP is a recommendation to include additional trips or Safety Instrumented Systems (SIS) in the design. Kletz said that this occurs because the HAZOP team perceives this to be a cheap and effective solution, but he pointed out that this was not the case [2].

---

### Text Box 1.35

Many are solved that way but not as cheaply as we usually think. Testing and maintaining an extra trip costs about as much as purchase and installation, even after discounting. Trips cost twice what we think. In addition, a considerable management effort is needed to make sure that the testing is carried out. Procedures can corrode faster than steel and vanish without trace once managers lose interest. Instead of installing extra trips, it is better to carry out systematic studies earlier in design to find ways of avoiding the hazards.

---

Another problem is that people assume that automated systems don't fail. Kletz pointed out that this is not correct [2].

---

### Text Box 1.36

Always consider the failure of automatic equipment as a possible cause of the deviations. For example, no flow may be due to a trip or controller failing to open a valve or closing it at the wrong time.

In fact he recommended that any software used as control needs to be subject to HAZOP [2].

---

**Text Box 1.37**

On the contrary, it is vital to HAZOP the instructions given to the computer (the applications software) as, unlike operators, the computer cannot recognise and query instructions that are ambiguous, lack precision or are simply wrong. Also, software engineers are rarely also chemical engineers and can easily misunderstand the chemical engineer's requirements.

---

Another very common issue is that alarms are often specified as safeguards in a HAZOP. This has, in part, contributed to a very serious problem in many industries including process, where alarm systems fail in their key objectives. This is because operators receive far too many nuisance alarms that do not require a response and are overwhelmed by floods of alarms that they have no chance of understanding whenever a system experiences a malfunction or upset. Many companies have identified this issue and are working hard to resolve it [9]. This often includes deletion or modifications to alarms. This can mean that assumptions made during the HAZOP are no longer valid and may call into question the underlying basis of the safeguards assumed in the study.

### 1.3.2.3 *Overlooking knock-on effects*

To be effective a HAZOP must not only identify issues, but it also should involve some consideration of potential solutions. However, there is a balance to strike because there is unlikely to be sufficient time to fully evaluate any proposed design change. Kletz identified that a failure to consider potential knock-on effects is one particular concern [2].

---

**Text Box 1.38**

When a change in design (or operating conditions) is made during a HAZOP, it may have effects elsewhere in the plant, including the sections already studied.

For example, during a HAZOP the team decided to connect an alternative cooling water supply to a heat exchanger. The original water supply was clean but the alternative was contaminated, and so the team had to change the grade of steel used for the heat exchanger and connecting lines. It also had to consider the effects of reverse flow in the original lines.

---

The way actions are worded can have a significant effect on how they are implemented. If they are too prescriptive it is possible that they are implemented without anyone taking the time to fully evaluate the intended and unintended consequences. However, high level actions to 'consider' a change are often closed quickly by someone saying that they have considered the issue and decided that no change was required.

## 1.3.3 Lack of follow-up

Of course a HAZOP does nothing on its own to improve safety. It is the actions that result that have this effect.

It is always reasonable for actions from HAZOP (or any study) to be rejected but they should not be ignored. In most cases, a rejection should be justified because another action is identified that can be demonstrated to be more effective. Whilst rejecting an action and doing nothing may be acceptable it should be clearly explained why the risks are considered to be acceptable and ALARP. "It is worse for the company to have a HAZOP and ignore the recommendations than not to have a HAZOP at all" [15].

One practice to be particularly aware of is for actions to be postponed to a later stage in the project [15]. This results in the original action being closed because the people currently involved in the project feel that they are unable to implement a solution. Such an action should only be closed if it has been transferred to another system (e.g. risk register) to ensure that it receives the proper attention at a later time. Alternatively, the action should be 'deferred' with details for when it needs to be addressed.

One problem with making HAZOP a mandatory activity, either as part of a project or as a planned retrospective review, is that it can become perceived as a 'tick-box exercise' that people think needs to be done quickly and easily, with the aim of discovering as few issues as possible. This can be partly avoided by ensuring that clear procedures are in place that describe how the HAZOP has to be performed but its importance still needs to be emphasised. Making a senior manager responsible for the final 'sign off' of the HAZOP report will focus their attention [14]. The responsibility in this regard is to confirm that all actions have been cleared or formally accepted as exceptions and that the HAZOP process has been carried out according to company guidelines by competent people.

### 1.3.4 Accidents that probably would *not have* been avoided by HAZOP

Kletz was quite clear that HAZOP cannot prevent every accident [2].

---

**Text Box 1.39**

They will not for at least three reasons: (a) Being human, we will not spot everything that might go wrong; (b) People will not always act in the way we have assumed they will; and (c) HAZOP cannot, by its nature, prevent mechanical accidents, such as people bumping into equipment which has been badly located.

---

Kletz highlighted that the HAZOP team have to assume that a system will be operated and maintained correctly [2].

---

**Text Box 1.40**

They assume that the general level of management is competent, that the plant will be operated and maintained in the manner assumed by the design team and in accordance with good management and engineering practice. In particular they assume that protective systems will be tested regularly and repaired promptly when necessary.

If these assumptions are not true then HAZOPs are a waste of time. It is no use identifying hazards or estimating their probability if no-one wants to do anything about them; it is no use installing trips and alarms if no-one is going to use or maintain them. The time spent on HAZOP would be better spent on bringing the safety consciousness of employees and management up to standard.

### 1.3.4.1 Piper Alpha

The tragedy at Piper Alpha that resulted in the death of 167 people when the offshore oil platform caught fire occurred because hydrocarbon leaked from the location where a Pressure Safety Valve (PSV) had been removed for maintenance. Whilst a HAZOP would have identified the PSV as a safeguard against overpressurisation of pipework it would have assumed that it would always be in place whenever hydrocarbon was present.

At its simplest level Piper Alpha occurred due to a leak of hydrocarbon from a pipework joint. It did not particularly matter which joint leaked. The fundamental failures were due to poor management of maintenance and isolations. Whilst HAZOP can highlight the importance of these processes at a general level it cannot consider specific scenarios.

Nonetheless, had a HAZOP had been performed on Piper Alpha, it may have identified the need for a HAZOP action to consider ways to mitigate the potential consequences of a vapour cloud explosion, as explained by Kletz as follows [12].

---

### Text Box 1.41

The overpressure from the explosion in the condensate area blew down the fire wall separating it from the section of plant containing equipment to extract crude oil. Containment was breached and there was an immediate and large oil fire. It was that fire and the resultant engulfing of the platform in thick smoke which so hampered attempts to escape.

Evidently, in the design stage of the platform, it had been judged that a fire could occur due to a loss of containment in the condensate section. The probability of that happening was sufficiently likely that fire walls were installed between the various sections of the unit. It is difficult to imagine that, in the light of that assessment, an explosion in the condensate area was any less likely. Nevertheless, Piper Alpha had no explosion walls either side of the condensate area. However, the more crucial deficiency was that, at the design stage, there had not been a systematic and, where necessary, quantified assessment of the major hazards which could occur on the platform. No hazard and operability review had been made of the design.

---

### 1.3.4.2 Buncefield

The fire and explosion at the Buncefield oil storage depot occurred because a tank used to store gasoline was overfilled. This occurred due to operational errors but level instrumentation that would have triggered an alarm when the tank reached a certain level was not functional.

A HAZOP would have identified a high level alarm as a safeguard against this accident, and it would have been a reasonable outcome of the study. But the HAZOP team have to assume that, once fitted, the instrumentation would have been maintained in an operational condition. The study could have been used to highlight the importance of the safeguard, but again this could only be at a general level because it would apply to a wide range of instrumentation associated with alarms and trips.

Other types of study such as LOPA may have provided a better insight into the adequacy of instrumentation that only caused an alarm without any trip function, but this would still assume that all devices were maintained appropriately.

Alternatively, an inherently safer design study may have identified options to eliminate the potential for the huge vapour cloud explosions by considering alternative options higher on the hierarchy of risk reduction techniques, rather than relying on additional layer of protection like level switches and valves which are all prone to failure. The consideration of inherently safer design at Buncefield is included in the Inherent Safety section of this book.

## 1.4 Practical application

The purpose of this chapter has been to share a range of viewpoints of HAZOP based on what Kletz wrote about and more recent experience by others. This final section provides some additional practical guidance for anyone who may be organising or taking part in a study.

### 1.4.1 The how and when of HAZOP

Performing a HAZOP requires a fair degree of planning and organisation. Kletz gave us some real insight into some of the issues that need to be addressed in doing this.

#### 1.4.1.1 Involve the right people

Kletz pointed out that HAZOP does not have magical properties and it totally relies on the involvement of people [2].

---

**Text Box 1.42**

Techniques discover nothing; only people discover hazards; techniques can help them do so. HAZOP helps people apply their knowledge and experience, and much of that comes from learning and remembering the lessons of the past.

---

Kletz discussed whether a HAZOP team should remain the same throughout a study. His view was that changes could occur, especially during longer studies as this would avoid too much HAZOP fatigue. Also, allowing different people to attend meant that the level of understanding and insight into the design, hazards, and potential operational issues would be distributed more widely through a team, which can be beneficial.

Kletz identified that some issues require expert input from people who may not be required to attend a HAZOP for the whole time [2].

---

**Text Box 1.43**

On the contrary, if a plant has specific problems (for example, if corrosion is serious), an expert in these problems should join the team. But do not have experts sitting in the meeting just in case their advice is needed.

---

Experts can always be on-call. The wide availability of tele- and video conferencing facilities means that they do not need to be local to have an input.

### 1.4.1.2 *Don't leave it to contractors*

One concern is for projects where the bulk of the design is performed by third parties. Often the completion of a HAZOP is part of their contract and that can lead to the impression that they should do this alone. Kletz highlighted that this would not be a good idea [2].

---

**Text Box 1.44**

Companies have been known to say to a design contractor, "We are under-staffed and you are the experts, so why don't you do the HAZOP for us?"

The client should be involved as well as the contractor because the client will have to operate the plant. The HAZOP gives the client's staff an understanding of the reasons for various design features and helps them write the operating instructions. Even if the client's staff know little to start with about the problems specific to the particular process, they will be able to apply general chemical engineering and scientific knowledge as well as common sense knowledge. Writing in a different context. Pegram says.[f] "... The only effective team is one that owns the problem. The team must therefore comprise the individuals who are responsible for implementing the results of the study, not an external group of experts." The actions agreed at a HAZOP include changes in procedures as well as changes to equipment and, while the contractor is responsible for the latter, the client is responsible for the former.

---

### 1.4.1.3 *Getting started*

Some people feel that they do not need to do HAZOP because they only ever carry out small projects where they (incorrectly) feel that it does not apply. Not only does this mean that the small projects are not given the scrutiny which they require, but it can mean that the company and people are not prepared when a large project comes along. Kletz summed up his views on this [2].

---

**Text Box 1.45**

On the contrary, we should start small by applying HAZOP to one or two projects and see if we find it useful. If we do, we can gradually increase our effort until we are HAZOPing all new plants and as many old ones as resources allow.

---

### 1.4.1.4 *Plant modifications should be HAZOPed*

Kletz had concerns that people sometimes feel HAZOP is only intended for new projects and so does not get used for modifications. He summed up his main concern as follows [2].

---

**Text Box 1.46**

A modification that has not been thoroughly thought through can result in a chain of further modifications during the subsequent months, possibly in distant parts of the plant.

---

[f]N. Pegram. The Chemical Engineer Number 482. 1990.

He gave the following further explanation [2].

---

### Text Box 1.47

Many people believe that HAZOP is unsuitable for small modifications because it is difficult to assemble a team every time we wish to instal a new valve or sample point or raise the operating temperature. However, many accidents have occurred because modifications had unforeseen and unpleasant side-effects. If proposals are not 'HAZOPed', therefore, they should still be thoroughly probed before they are authorised.

Do not overlook the following modifications:

- Temporary modifications as well as permanent ones;
- Start-up modifications as well as those on established plants;
- Cheap modifications as well as expensive ones; and
- Modifications to procedures, process materials or operating conditions, as well as modifications to equipment.

---

The requirement to consider a HAZOP should be integral in any Management of Change procedure. It should be very clear in which circumstances it should be mandatory, recommended, or not required.

#### 1.4.1.5 'Reasonably practicable' applies

One concern is that companies are obliged to implement actions arising from HAZOP, even if they are very expensive. Kletz said that this was not the necessarily the case [2].

---

### Text Box 1.48

HAZOP uncovers problems and ways of removing them. If the problems are very expensive to remove and the risks are not intolerable (in the legal phrase, if removing them is not 'reasonably practicable'), we do not have to remove them. But it is wrong to close our eyes in case we do not like what we see.

---

What is clear is that making a case for ALARP requires far more than simply rejecting an action because it is expensive. Any rejection of an action should have a fully documented justification, primarily focussed on why the risk reduction achieved is not considered sufficient. Ideally, an alternative action should be proposed that addresses the issue raised in the HAZOP and which is more effective and/or less expensive.

#### 1.4.1.6 Automated HAZOP

Software for recording HAZOP inputs and outputs is now ubiquitous, but people would like to take this further. Automating the HAZOP process has been researched by a number of groups for many years, but none have succeeded in developing a usable system, perhaps because computers cannot replace the imagination and judgement of a knowledgeable and experienced team. Kletz had some involvement in this and shared his views [2].

**Text Box 1.49**

Computers are now widely used for recording results and reminding teams of common causes of deviations. Programmes have also been developed, but so far little used, for producing a list of technical problems for consideration. A computer is highly unlikely to be able to identify the problems that arise out of interactions, or failures to interact, between people.

#### 1.4.1.7 Other types of HAZOP

It has been stated that HAZOP is only one method of many that can be used for hazard identification studies. There are some variations on the method that have been developed for specific purposes. Some of the more well known include:

- Batch HAZOP, for batch processes, where the guidewords should be augmented to include phrases such as 'early', 'late', 'before', 'after', etc.;
- Human or procedures HAZOP—focussed on human errors, their consequences and controls; and
- Software or control HAZOP (CHAZOP)—identification of weaknesses and errors in software.

Studies using these methods can follow very similar approaches taken to standard HAZOP, but will require team leaders and/or team members with the appropriate subject knowledge.

### 1.4.2 Specific pointers for process HAZOP

Kletz provided us with a lot of things to look out for when carrying out HAZOP in the process industry.

#### 1.4.2.1 HAZOP for vessels

Kletz suggested that vessels sometimes need special consideration during HAZOP [2].

**Text Box 1.50**

When all the lines leading into a vessel have been studied, the guide word OTHER THAN is applied to the vessel. It is not essential to apply the other guide words to this item as any problems should come to light when the inlet and exit lines are studied. However, to reduce the chance that something is missed, the guide words should be applied to any operation carried out in the vessel. For example, if settling takes place we ask if it is possible to have no settling, reverse settling (that is, mixing), more settling or less settling, and similarly for stirring, heating, cooling and any other operations.

#### 1.4.2.2 Intermediate storage

Kletz's focus on inherent safety led to him saying that "you should have grey hairs, not intermediate storage." Nevertheless, it is often required and it should receive special attention during a HAZOP [2].

**Text Box 1.51**

Pay special attention to intermediate storage vessels. As a rule, no change is supposed to take place there except emptying or filling but changes in temperature or composition may take place, particularly when the contents are allowed to stand for longer than usual.

### 1.4.2.3 Small branches

Kletz highlighted the potential for HAZOPs to focus on larger items, which sometimes means that apparently less significant items such as small branches can be missed [2].

**Text Box 1.52**

Do not overlook small branches which may not have been given a line number. For example, a tank was fitted with a tundish so that it could be dosed with stabilising chemicals. The effects of adding too much or too little additive (or the wrong additive, or adding it at the wrong time) should obviously be considered during HAZOP but might be overlooked if the team studied only lines with line numbers. (On the other hand, they might have picked it up by considering operations taking place inside a vessel, another example of the way in which HAZOP often gives us a second chance.)

### 1.4.2.4 Ask about design features

Kletz suggested that the HAZOP was a good time to ask about a range of design features that are not always directly associated with the process [2].

**Text Box 1.53**

The HAZOP also provides an opportunity to check that a number of detailed points have been considered during design. The team should ask:

- What types of gasket have been used? Should spiral wound ones be used? Has the number of types been kept to a minimum? (The more types we use, the greater the chance that the wrong sort will be used.)
- Has the number of types of nuts and bolts, been kept to a minimum?
- Are the valves used of a type, such as rising spindle valves, whose position can be seen at a glance? If ball valves or cocks are used, can the handles be fitted in the wrong position?
- Are spectacle plates installed whenever regular slip-plating (blinding) of a joint (for maintenance or to prevent contamination) is foreseen?
- Access is normally considered later in design, when a model of the plant (real or on computer) is available, but the HAZOP team should note any points that need special attention, for example valves that will have to be operated frequently or in an emergency, and should therefore be easy to reach.

## 1.4.3 Summing up

Kletz shared the following about the potential power of HAZOP [2].

---

### Text Box 1.54

Carling[g] has described the effects of using HAZOP in his company. The benefits went far beyond a simple list of recommendations for a safer plant. The interaction between team members brought about a profound change in individual and departmental attitudes. Staff began to seek one another out to discuss possible consequences of proposed changes, problems were discussed more openly, departmental rivalries and barriers receded. The dangers of working in isolation and the consequences of ill-judged and hasty actions became better appreciated. Knowledge, ideas and experience became shared more fully to the benefit of the individual and the company.

Carling's company adopted HAZOP after experiencing several serious incidents. Buzzelli writes[h]: "For an industry so proud of its technical safety achievement it is humbling to have to admit that most of our significant safety improvements were developed in response to plant accidents."

It does not have to be so. HAZOP provides us with a lantern on the bow, a way of seeing hazards before they wreck our ship.

---

A publication that reviewed 40 different methods of hazard identification [9] summarised the advantages of HAZOP as follows.

- It is a systematic and comprehensive technique. A detailed plan for performing the technique is available which systematically applies guidewords and parameters to all the pipes and vessels in the process.
- It examines the consequences of the failure. Thought should be given by the assessment team to the consequences of the deviations identified. This aids in the production of recommendations for methods to minimise or mitigate the hazard.

The same study summarised the disadvantages of HAZOP as follows.

- It is time consuming and expensive. Most plants contain a large number of pipes and vessels each of which need to be examined by the application of the various guidewords and parameters.
- It requires detailed design drawing to perform the full study. To fully perform the study the process has to be designed to such a level that all the pipes and vessels are detailed with their operating conditions and control instrumentation.
- Additional guidewords are required for unusual hazards. For specific dangers that will not be covered by the general guidewords, further words (such as radiation for the nuclear industry) will need to be applied.

---

[g]N. Carling. HAZOP study of BAPCO's FCCU complex. American Petroleum Institute Committee on Safety and Fire Protection Meeting, Denver, Colorado. 1986.

[h]D. Buzzelli. Plant/Operations Progress, 9 (3) 145. 1990.

- It requires experienced practitioners. Experienced team members are required to identify all possible causes and consequences of the deviations, as well as producing realistic recommendations.
- It focuses on one-event causes of deviation only. Only the hazards associated with single deviations can be studied. Hazards that are caused by two or more separate deviations cannot be identified by the technique.

Another study [13] concluded that HAZOP is very beneficial at assuring the 'conception, configuration and specification' of a system, which leads to improvements in specifications for fabrication, commissioning, operating, and maintaining the plant. But it has many side benefits, including serving as a source of reference for aiding decisions during the system's life, particularly when making modifications and for educating the people involved in a way that helps to improve future design and eliminate cross-discipline conflicts.

HAZOP conducted at design provides the opportunity to avoid high probability, low consequences events, whilst also ensuring suitable protection is in place against the low-probability, high-consequence events. HAZOPs conducted retrospectively have the advantage of experience that gives a greater insight into problems and the likelihood of escalation into incidents and possibly accidents [13].

HAZOP is one of the few methods for which formal guidance exists. It should not be seen as the only option for identifying hazards, with the chosen method depending on the application and available resources [9].

## References

[1] T. Kletz, By Accident … a Life Preventing Them in Industry, PFV Publications, 2000.

[2] T. Kletz, HAZOP and HAZAN, fourth ed., IChemE, 1999.

[3] B. Tyler, HAZOP study training from the 1970s to today, Process Saf. Environ. Prot. 90 (2012) 419–423.

[4] G. Eggett, S. Whitey, The Leadership and Management of HAZOP Study, IChemE Course, Leeds, 2008 20–22 October.

[5] F. Crawley, B. Tyler, HAZOP Guide to Best Practice, third ed., Elsevier, 2015.

[6] International Standards Organisation, ISO 17776-2000: Petroleum and Natural Gas Industries—Offshore Production Installations—Major Accident Hazard Management During the Design of New Installations, (2000).

[7] G. Ellis, How to Speed Up Retrospective HAZOPs Whilst Achieving Effective Risk Reduction, HazardEx, 2018.

[8] C. Crowley, HAZOPS are not the only fruit, Loss Prev. Bull. 237 (2014) 17–20.

[9] M. Glossop, A. Ioannides, J. Gould, Review of Hazard Identification Techniques, Health and Safety Laboratory, 2000.

[10] Health and Safety Executive, The Explosion and Fires at the Texaco Refinery, Milford Haven. 24th July 1994, http://www.hse.gov.uk/comah/sragtech/casetexaco94.htm, 1994. Accessed October 2019.

[11] A. Musthafa, Managing risk in major maintenance—a case study on fire and explosions incidents in the process industry, Loss Prev. Bull. 268 (2019) 11–14.

[12] T. Kletz, Learning from Accidents, third ed., Butterworth-Heinemann, 2001.

[13] A. Rushton, Quality assurance of HAZOP. Health and Safety Executive offshore technology report OTO 96 002, (1996).

[14] C. Feltoe, HAZOP failure, Loss Prev. Bull. 232 (2013) 19–22.

[15] J.R. Taylor, Lessons learned from forty years of HAZOP, Loss Prev. Bull. 227 (2012) 20–27.

[16] P. Eames, Hazard identification: planning for success, Chem. Eng. (2019) https://www.thechemicalengineer.com/features/hazard-identification-planning-for-success/.

[17] U.S. Chemical Safety and Hazard Investigation Board, Key lessons for preventing hydraulic shock in industrial refrigeration systems. Anhydrous ammonia release at Millard Refrigerated Services, Inc., CSB Saf. Bull. (2015).

[18] T. Fishwick, Rupture of a heat exchanger at a refinery causes fatalities, Loss Prev. Bull. 228 (2012) 11–14.

[19] K. Kidam, M. Hurme, Method for identifying contributors to chemical process accidents, Process Saf. Environ. Prot. 91 (2013) 367–377.

[20] A. Younes, The good, and bad, carrying out retrospective HAZOP studies for a large scale off-shore facilities, in: IChemE Hazards 26, 2016.

# Hazan (Quantified risk assessment)

## 2.1 Introduction

Kletz makes clear that 'Hazan', a contraction of hazard analysis, in his nomenclature is equivalent to the more commonly used QRA (quantified risk assessment). He addressed Hazan at length and in detail in only one of his many books, from which all of the quotations in the text boxes are taken [1].

---

**Text Box 2.1**

Why do we want to apply numerical methods to safety problems?

The horizontal axis [in Fig. 2.1 below] shows expenditure on safety over and above that necessary for a workable plant, and the vertical axis shows the money we get back in return. In the left-hand area safety is good business—by spending money on safety, apart from preventing injuries, our plants blow up or burn down less often and we make more profit.

In the next area safety is poor business—we get some money back for our safety expenditure but not as much as we would get by investing our money in other ways.

If we go on spending money on safety we move into the third area where safety is bad business but good humanity—money is spent so that people do not get hurt and we do not expect to get any material profit back in return—and finally into the fourth area where we are spending so much on safety that we go out of business…. We have to decide where to draw the line. Usually this is a qualitative judgement but it is often possible to make it quantitative. The methods for doing so are called, in this book, hazard analysis or Hazan. Other names are risk analysis, quantitative risk assessment (QRA) and probabilistic risk assessment (PRA).

---

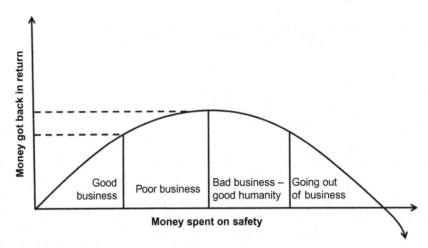

**FIG. 2.1**   The effects of increasing expenditure on safety.

The Hazan techniques described by Kletz

---

### Text Box 2.2

   assume that the general level of management is competent, that the plant will be operated and maintained in the manner assumed by the design team and in accordance with good management and engineering practice. In particular, they assume that protective systems will be tested regularly and repaired promptly when necessary... It is no use identifying hazards or estimating their probability if no-one wants to do anything about them; it is no use installing trips and alarms if no-one is going to use or maintain them.

---

   Indeed, not doing anything with the results of the assessment is one of the main pitfalls in the way risk assessment is used in industry [2]. Arguably an even worse pitfall is using a risk assessment to justify a decision that has already been made; though such a scheme has sometimes been disguised as commendable diligence. Carrying out a detailed quantified risk assessment should not be considered without first considering whether any relevant good practice is applicable. For example, the UK's Health and Safety Executive (HSE) Railway Inspectorate issued a Prohibition Notice against a railway operator requiring them to include a specific safety procedure that involved staff carrying out physical checks on trains before carrying out a specific manoeuvre. The railway operator appealed on the grounds that it was not reasonably practicable for them to physically check trains, because the cost was grossly disproportionate to the benefit—as they purported to demonstrate with their QRA and Cost–Benefit Analysis. However, they ignored relevant industry good practice in relation to the manoeuvre, not to mention their own adherence to the same good practice on another part of their operation, effectively demonstrating that the action was reasonably practicable [2] (see Chapter 4 for further information about requirements to achieve risks that are ALARP—As Low As Reasonably Practicable).

## 2.1.1 Benefits of Hazan

Where Hazan/QRA is appropriate (i.e. complex, high hazard systems), not doing it can result in effort and expense being lavished on the wrong targets with too little focus on the main risks:

A very important message is that the numerical result from a QRA is not the only benefit from performing an assessment. Kletz highlighted that we should be using the whole process to improve our understanding of risk and how it can be controlled.

---

**Text Box 2.3**

Hazan does more than tell us the size of a risk. Especially when fault trees are used, it shows how the hazard arises, which contributing factors are the most important and which are the most effective ways of reducing the risk.... If we deal with each problem as it arises, the end result may be the opposite of that intended. This is common in politics and can also occur in engineering. It can result in massive expenditure on preventing a repetition of the last accident while greater risks, which have not so far caused injury, are unrecognised and ignored.

---

Kletz used the following example to indicate where failure to quantify risks can mean opportunities to apply more effective controls can be missed.

---

**Text Box 2.4**

In 1987, 31 people were killed and many injured by a fire in King's Cross Underground railway station in London. The official report made 157 recommendations. However, ... it failed to use QRA or cost-benefit analysis and made recommendations that would produce little benefit ... Intuitively, [replacing the wood in every escalator with metal] had seemed a good idea, but calculations showed that this would reduce the probability of a serious escalator fire from once in six years to once in nine, while installation of sophisticated sensors and automatic sprinklers would reduce the probability to once in a thousand years.

---

At first glance, this example would seem to run counter to Kletz's forward thinking on inherent safety (see Chapter 4 of this book). The inherent safety approach would be to reduce reliance on engineered controls (e.g. sensors and automatic sprinklers), and instead replace hazardous items (e.g. wooden escalators) with less hazardous ones (e.g. metal escalators). However, the point he makes here is on prioritisation: remove the greatest risks first. The QRA had shown the wooden treads to make a very small contribution to the total risk of escalator fires. The wooden escalators on the London Underground network were not, in fact, replaced as a matter of urgency, the exercise being completed only in 2014.

The cure can even be worse than the disease. There have been numerous incidents where contaminated firefighting water from chemical warehouse fires has entered waterways causing massive fish kills, in circumstances where the chemicals themselves would not have reached the waterway [3]. Recognising this issue and incorporating it into risk assessment can result in a much better understanding of what controls are required. It was thanks to a risk assessment that a vital aquifer was protected from contamination when the Sherwin-Williams paint warehouse was allowed to burn down. The risk to the aquifer due to fire-water runoff was assessed to outweigh that from the continuing smoke plume, which could have been reduced if firefighting had continued [4]. A method has recently been published for assessing toxic risks from smoke plumes emitted by warehouse fires, to support controlled burn-down decisions in planning for and dealing with large fires [5].

## 2.1.2 Process vs personal safety

The use of QRA is especially appropriate for analysing low-probability, high-consequence process risks. Kletz pointed out that measures of safety performance at the other end of the risk spectrum do not inform us about these risks.

### Text Box 2.5

The most widely used measure of safety, the lost-time accident rate, is deeply flawed. All lost-time accidents are not comparable: better that 100 people are absent for a few days with minor bruises than one person is blinded or paralysed. In addition, lost-time accidents are now so few in most companies that their rate measures luck and the willingness of injured people to continue at work. A low lost-time accident rate does not indicate that technical safety problems are under good control.

A misguided focus on the lost-time accident rate as a surrogate for process safety has been identified in official investigations as a significant contributor to several major accidents, such as the BP Texas City refinery disaster (see Section 3 of Appendix) in 2005, which caused 15 deaths and more than 170 injuries [6].

## 2.1.3 The stages of Hazan

Kletz defined three stages:

### Text Box 2.6

Every Hazan, however simple, consists of three steps:

**(i)** Estimating how often the incident will occur.
**(ii)** Estimating the consequences to:

- employees;
- the public and the environment;
- plant and profits.

...

(iii) Comparing the results of (i) and (ii) with a target or criterion in order to decide whether or not action to reduce the probability of occurrence or minimise the consequences is desirable, or whether the hazard can be ignored, at least for the time being.

...

In brief, the stages in hazard analysis are:

(i) How often?
(ii) How big?
(iii) So what?

Kletz explains his reasons for focusing on (iii) and (i)—in that order—whilst more or less omitting (ii):

## Text Box 2.7

Discussion of (ii - how big?) is not attempted. The methods used differ for each kind of hazard – fires, explosions and releases of toxic gas – and the number of calculation methods available is enormous... Refer to specialist textbooks or to Lees.

...

Many writers are reluctant to discuss step (iii), but it is little use knowing a plant will blow up once in 1000 years with a 50% chance that someone will be killed, unless we can use this information to help us decide whether we should reduce the probability (or protect people from the consequences) or whether the risk is so small, compared with all the other risks around us, that we should ignore it and devote our attention to bigger risks. For this reason, step (iii), setting a target criterion, is discussed before step (i), estimating how often an incident will occur.

## 2.2 Risk criteria

Kletz forcefully argues in defence of risk criteria to inform and motivate safety management:

## Text Box 2.8

When injury is unlikely we can compare the annual cost of preventing an accident with the average annual cost of the accident... This method could be used for all accidents if we could put a value on injuries and life, but there is no generally agreed figure for them.

Various ways have been suggested for estimating the cost of saving a life... The range [of prices actually paid] is enormous. Doctors can save lives for a few thousands or tens of thousands of pounds per life saved and road engineers for a few hundred thousands per life saved, while industry spends millions and the nuclear industry tens of millions (even more according to some estimates) per life saved.

What value then should we use in cost-benefit calculations? I suggest the typical value for the particular industry or activity (such as the chemical industry or road safety) in which we are engaged. Society as a whole might benefit if the chemical or nuclear industries spent less on safety and the money saved was given to road engineers or to doctors, but there is no social mechanism for making the transfer…

After Flixborough a BBC reporter, standing in front of the plant, described the explosion as 'the price of nylon'. Many people must have wondered if it is worth taking risks with men's lives so that we can have better shirts and underclothes. However, in our climate we have to wear something. How does the 'fatal accident content' of wool or cotton clothes compare with that of clothes made from synthetic fibres? The former is certainly higher. The price of any article is the price of the labour used to make it, capital costs being other people's labour. Agriculture is a high accident industry; so there will be more fatal accidents in wool or cotton shirts than in nylon shirts.

To many people the calculations of this chapter… may seem cold-blooded or even callous. Safety, like everything else, can be bought at a price. The more we spend on safety, the less we have with which to fight poverty and disease or to spend on those goods and services which make life worth living, for ourselves and others. Whatever money we make available for safety we should spend in such a way that it produces the maximum benefit… Those who make the sort of calculations described in this chapter, far from being cold-blooded or callous, are the most effective humanitarians, as they allocate the resources available in a way which will produce the maximum benefit to their fellow men.

Kletz mentions a common situation in which risk criteria can be set without estimating the risk to life:

## Text Box 2.9

When we are making a change it is often sufficient to say that the new design must be as safe as, preferably safer than, that which has been generally accepted without complaint. For example:

- If trips are used instead of relief valves they should have a probability of failure 10 times lower.
- If equipment which might cause ignition is introduced into a Zone 2 area it should be no more likely to spark than the electrical equipment already there.
- A new form of transport should be no more hazardous, preferably less hazardous, than the old form.

Again, in simple situations, "a sort of intuitive Hazan" is all that is needed to guide actions that will ensure risk criteria are met:

For example, in fixing the height of handrails round a place of work, the law does not ask us to compare the cost of fitting them with the value of the lives of the people who would otherwise fall off. It fixes a height for the handrails (36–45 in.) … [such that] the chance of falling over them, while not zero, is so small that we are justified in ignoring it.

### 2.2.1 Societal risk criteria

We often consider risk in terms of the impact to people individually—the likelihood that someone will be harmed. However, this does not provide a meaningful understanding of the risk of an event that can harm many people at the same time. This is defined as societal risk.

Kletz controversially addresses societal risk criteria by weighing the arguments for prioritising either of two cases.

---

**Text Box 2.10**

**(A)** One person is killed every year for 100 years.
**(B)** 100 people are killed once in 100 years.

---

He concludes that:

---

**Text Box 2.11**

The simplest and fairest view seems to be to give equal priority to the prevention of (A) and (B)—we're just as dead in case (A) as in case (B).

---

Kletz's view is an extreme one, but his supporting reasoning is persuasive. He counters the customary use of formulae to account for scale aversion (see later) on both practical and ethical grounds.

---

**Text Box 2.12**

However, these formulae are quite arbitrary and if we divide the hazard rate [designated tolerable] by $N^2$, or even N [the number of people killed per incident], we may get such low hazard rates that they are impossible to achieve.

…

If we give priority to the prevention of (B) we are taking resources away from the prevention of (A) and, in effect, saying to the people who will be killed one at a time that we consider their deaths as less important than others.

---

As Kletz points out, risk criteria are not the concern of experts alone, but a matter on which the public—and especially those exposed to the risk—have a right to comment. In a democracy, risk criteria must be acceptable to the public. The problem is that public opinion on this matter is so difficult to pin down because it is very subjective and can be easily influenced by media coverage. According to HSE [7], evidence exists both for and against a tendency for the public to want greater protection where consequences are high (described as 'scale aversion') and, even where such a tendency is evident, it is not consistent but varies with numerous factors many of which are themselves subjective. Whilst scale may be a contributing factor to how people perceive risk both before and after an accident, it is rarely the dominant factor. HSE concludes that it is neither practical nor sensible to attempt to formulate scale aversion in mathematical terms.

Societal risk assessment remains conceptually problematic. For example, how should we classify the overall societal risk where risks of scenarios potentially affecting different numbers of people differ widely in their tolerabilities? Again, it can be difficult to compare societal risks for different sites, and sometimes even for different risk reduction options at a given site.

Partly for these reasons, efforts have been made by HSE and others to organise or manipulate societal risk data sets to generate a single characteristic number. Unfortunately, this is to trade transparency for convenience: such risk indices are convenient for comparisons, but their meaning can be far harder to discern than that of the originating data. In the case of the Scaled Risk Integral (SRI), even the dimensions are baffling (no matter that they are logically derived) [8]:

$$(\text{people}^2 + \text{people})/\text{hectare}$$

Probably the simplest and most natural numerical index is obtained by multiplying the frequency of each component incident by the number of fatalities it would cause and summing the results over all components. The result is the calculated annual fatality rate or Potential Loss of Life (PLL) from all accidents at the major hazard facility.

PLL and SRI are polar opposites in their treatment of the public's putative scale aversion: PLL ignores it completely, whilst SRI vastly exaggerates it.

Much of the guidance from the UK's HSE cites its landmark publication "Reducing risks, protecting people – HSE's decision making process" [9], affectionately known as R2P2, which recommends an ostensibly scale-neutral approach for societal risk criteria. R2P2 locates at the tolerability limit, and thus effectively equates, accidents causing 100 or more deaths at a frequency of $10^{-4}$ per year with those causing 10 or more at $10^{-5}$ per year, those causing 1000 or more at $10^{-3}$ per year and so on pro rata.

Despite appearances, however, this approach is in fact far from scale neutral. The words 'or more' have been shown to conceal a surprising degree of scale aversion [10, 11]. Whilst the philosophical argument continues about the advisability of including scale aversion in setting societal risk criteria, in practice the available choice is between moderate and severe expressions of scale aversion.

Near the end of his section on societal risk criteria, Kletz gives us an important reminder.

---

**Text Box 2.13**

We should never decide that a risk is tolerable on the basis of [societal risk results] alone. We should also consider the people at greatest risk.

---

## 2.2.2 Risk criteria for accidents to the environment

Kletz looks briefly, with a somewhat jaundiced eye, at attempts to apply cost–benefit analysis to risks to the environment:

---

### Text Box 2.14

The [costs of prevention] are comparatively easy to estimate. Some of the costs of pollution can also be estimated... It is much more difficult to put a price on the intangibles, such as the aesthetic value of pleasant surroundings or the desire to preserve as much as possible of the natural world and the evidence of the past... People want the benefits and would rather not know the price, unaware that they are paying it. In a world in which many people are still suffering malnutrition and preventable disease, the value of some expenditure on improving the environment may be doubted. We should at least know what it is costing and what else could be done with the money.

It is also difficult to specify types of incident and frequencies that can be considered intolerable or broadly acceptable. A first attempt in that direction has been made by the UK Department of the Environment. It has listed 13 events that could constitute major environmental accidents. They include... death (or inability to reproduce) of 1% of any species. While loss of 1% of the world's population of, say, chimpanzees, may well be a major accident, it is difficult to feel the same about 1% of a species of beetle.

---

Kletz is right to point out the difficulties in attempting to set environmental risk criteria; however, it appears they are not insurmountable. The most useful approach, originating in an EU study [12] and summarised by Vince [4], has been to assign each accident scenario an environmental harm index (EHI), by comparing it with the reference accident, which is set as the minimum to qualify as a major accident to the environment (MATTE) in the circumstances. The reference accident is defined as the product of a reference extent (the length of a river or the area of an estuary or lake, found amongst the DETR tabulated values [13]), a reference measure of harm ($LC_{50}$ for the most vulnerable representative species—even if, *pace* Kletz, this happens to be a species of beetle), and a reference recovery time (5 years).[a]

Risk criteria, developed in key project workshops, were based on the following considerations:

- Consequence modelling of real accidents indicates that MATTEs typically have EHI values of at least 100.
- In respect of consequences for human health and safety, a frequency of $10^{-4}$ per year is regarded by HSE as being on the borderline of tolerability.
- Risk criteria schemes in common use typically have an ALARP zone[b] extending two orders of magnitude in both consequence and frequency dimensions.

---

[a] Suggested by information from substantial environmental accidents resulting in prosecutions.

[b] Where risks are tolerable if they have been brought As Low As Reasonably Practicable (ALARP).

– The gradient of the lines forming the zone boundaries (i.e. the rate at which tolerable frequencies decrease with increasing EHI) is the same as that used in other discussions in United Kingdom on risk criteria, namely −1. So, for example, doubling the calculated EHI of an accident halves its maximum tolerable frequency. (NB since we are here considering discrete scenarios, there is no feature corresponding to the 'or more' of societal risk criteria discussed earlier, and thus no scale aversion.)

After a long gap, the earlier work has been further developed by a working group of CDOIF (the UK Chemical and Downstream Oil Industry Forum), extending the methodology to include groundwater, terrestrial receptors, and the built environment, and breaking it into screening and QRA phases, amongst other enhancements [14].

## 2.3 Frequency/probability assessment

Kletz emphasises the primacy of past experience, as far as possible specific to the subject plant or a near equivalent, in determining how often an incident will occur.

---

**Text Box 2.15**

However, sometimes there is no past experience, either because the design is new or the incident has never happened, and in these cases we have to use synthetic methods.

---

### 2.3.1 Latent faults

Hazardous process plant is equipped with a variety of protective systems. Incidents occur or escalate when one or more of these systems fails to operate when needed.

Ultimately, it is the fraction of the time that a protective system is inactive that is of most concern. This is the probability that it will fail to operate when required and is known as the fractional dead time (fdt). This is calculated as a function of the demand rate (the frequency at which the protective system is called on to act), the frequency at which the protective system develops faults, and the frequency of testing the protective system.

Kletz highlighted how calculations lead us to learn about how systems operate and hence what we can do to improve reliability and safety to required levels. For example, recognising the difference between revealed and hidden (or unrevealed) failures allows us to understand which systems need to be tested and what those tests need to achieve.

---

**Text Box 2.16**

The failure of some equipment is obvious and is soon noticed by operators. Relief valves and trips, however, are normally not operating and their failures remain latent or unrevealed until a demand occurs. Hence we have to test them regularly to detect failures.

It is sometimes possible, by a change in design, to turn a hidden fault into a revealed one. For example, the failure of an alarm bell or hooter is hidden. If it fails, it is out of action until it is tested and repaired. We test frequently and accept a small chance that we may not know when an alarm occurs. If we want greater reliability, then instead of a bell that rings when an alarm is signalled we can have a device that sounds continually but becomes louder when there is an alarm. If the sound stops, we know something is wrong. Another example: failure of the front light on a bicycle is noticed at once; failure of the rear light is not. If the two lights are in series, failure of either is noticed (but then we have no lights at all).

If a trip is never tested, then after a few years the fractional dead time will be 1—that is, the trip will be 'dead', and the hazard rate will be the same as the demand rate.

Some companies test 'critical' trips and alarms but not 'noncritical' ones. If a trip or alarm is so unimportant that it does not need to be tested, it is probably not needed. If its failure rate is 0.5/year then after 4 years the probability that it will be in working order is less than 10%. (However, if an alarm is fitted to a control or indicating instrument, certain failures—such as a failure of the sensor—may be obvious to the operators and it will then be repaired.)

If the trip is tested yearly, then [assuming one demand per year] the hazard rate is only reduced from once/year with no trip to once in 5 years. If the trip is so unimportant that annual testing is sufficient, then the trip is probably not necessary.

---

The fdt decreases with decreasing test interval, but only up to a point. If system testing is carried out too frequently, the fdt begins to increase, because the time the system is inoperative because of testing and repair becomes comparable to the time it is inoperative because of faults. In addition, each test may be an opportunity for human error to leave the system in a fail-danger state. Notoriously, the Buncefield incident (see Section 14 of Appendix) [15] was precipitated by an independent high-level switch that was left effectively inoperable after testing (although this was due to more to a design fault than operator error). Given relevant quantitative information on the earlier issues, an optimum test interval can be calculated, at which fdt is a minimum.

The overall fdt (and thus the hazard rate) of protective systems installed in parallel, e.g. two (or more) high level trips, where only one system needs to be operative to prevent an incident, is of course lower than that of a single system. The precise value depends on the testing protocol, with staggered testing of the systems always better than simultaneous testing. When calculating the fdt of such systems, it is common, but unduly optimistic, to assume that the systems are independent (i.e. the failure of one system does not affect the probability of any of the others failing). In practice, this assumption is always to some degree unjustified; dependent failures are discussed below.

For protective systems in series, e.g. a relief valve and bursting disc (as often employed to prevent product leakage and prolong valve life), the overall fdt and hazard rate are higher than they would be with either system on its own. This is because, for an incident to occur, only one system needs to fail; conversely, to prevent an incident, every system needs to be available on demand. For example, relief valves and bursting discs are devices to protect plant from high pressure. If either fails to operate on demand (i.e. the disc does not burst or the valve does not lift), the protection will fail.

Kletz explained:

---

**Text Box 2.17**

If we connect in series many items of equipment each of which has high reliability—that is, a low fractional dead time—the overall system may be very unreliable. For example, if there are 10 items and each has an fdt of 0.05, the overall fdt will be about 0.4. For this reason, in high-risk situations, novel systems take a long time to become established and proven designs of aircraft and nuclear reactors continue to dominate the market.

---

Kletz warns against the common temptation to use protective systems for process control:

---

**Text Box 2.18**

If we increase the demand rate on a protective system we increase the failure rate. When more protective systems are added to a plant there may be a tendency for operators to increase the demand rate on them and if they do we may soon be back with the old failure rate… It is a useful exercise to calculate the hazard rates of our trip systems, from failure rates, demand rates and test intervals. We may find that to get an acceptable hazard rate we have to assume that nine out of ten deviations are spotted by operators before the trip operates. Do operators realise this? Do managers realise this?

---

Whilst from a safety perspective our main concern is the failure of systems to operate on demand, Kletz pointed out that production gives us a different perspective—with its own safety implications.

---

**Text Box 2.19**

As well as fail-danger faults, there are the so-called fail-safe faults or spurious trips—the protective equipment operates although there is no hazard. For example, a relief valve lifts light or a high level trip operates when the level is normal. I say 'so-called' because they may be unsafe in other ways; they may result in a discharge to atmosphere or an unnecessary shutdown of the plant, which may cause a leak. They give protective systems a bad name and make them unpopular with plant operators who may be tempted to bypass them.

---

As mentioned earlier, multiple protective systems in parallel will substantially reduce the overall fdt. However, the potential for spurious (fail-safe) trips increases with the number of installed protective systems. Voting systems are a useful compromise, the most common being a 2-out-of-3 arrangement, where the trip operates only if at least two out of the three installed measuring devices demand it.

Kletz highlighted the reason for using voting systems: production, not safety.

**Text Box 2.20**

Voting reduces the fail-safe or spurious trip rate and is used when spurious trips would upset production. It does not give increased safety. A 1-out-of-2 system is three times safer than a 2-out-of-3 system.

The above comparison represents a theoretical worst case. In general, the degradation in safety in going from a 1-out-of-2 to a 2-out-of-3 systems is less marked.

**Text Box 2.21**

Fail-danger faults remain hidden (latent, or unrevealed) until there is a test or demand. In practice [in a voting system], a single fault signal usually sounds an alarm and the fault is thereby disclosed. If this is the case, then instead of the test interval T the repair time should be used in the formula [for the hazard rate] (or, more precisely, the time from the alarm sounding to the completion of the repair).

## 2.3.2 Fault trees

Kletz provides a very brief introduction to fault tree analysis (FTA). The technique, originally developed in the 1960s in a military context [16], has found extensive use in the field of reliability, especially for the analysis of complex systems. A very simple set of symbols and rules based on reliability theory, Boolean algebra, and probability theory provides the mechanism for analysing very complex systems and complex relationships between hardware, software, and humans.

**Text Box 2.22**

Fault trees are widely used in Hazan to set down in a logical way the events leading to a hazardous occurrence [aka top event]. They allow us to see the various combinations of events that are needed and the various ways in which the chain of events can be broken. They allow us to calculate the probability of the hazardous event from other probabilities that are known.

In drawing a fault tree we start on the left [or the top]… We then work from left to right (or top to bottom) drawing in the various events that lead up to the top event. … In practice we stop drawing when we have data for the frequency of the events or the probability of the conditions on the right (or the bottom) of the tree. … Then we work back inserting numbers and estimate the frequency of the top event.

Amongst its benefits, the top–down, deductive approach in the construction of the fault tree, working backwards from the undesired effect to its potential causes, is a useful complement to the bottom–up approach of Hazop.

## Text Box 2.23

The points at which two[c] branches of the tree join are known as gates; they can be 'AND' or 'OR' gates. ... At school we were taught that AND means add. Remember that in drawing fault trees:

– OR means add;
– AND means multiply

Using a light-hearted illustration of a simple fault tree, in which he assesses the 'risk' of a free lunch under various circumstances, Kletz draws out several valuable lessons (Figs 2.2–2.4).

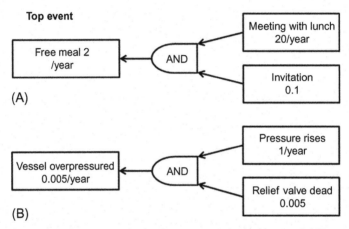

**FIG. 2.2**   Fault trees with 'AND' gates. Note that a *frequency* is multiplied by a *probability*.

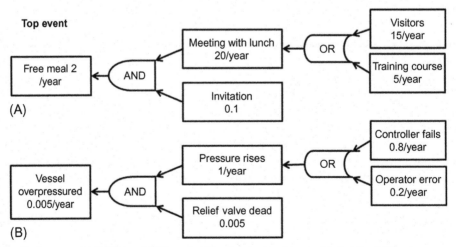

**FIG. 2.3**   Fault trees with 'AND' and 'OR' gates. Note that *frequencies* are added at the 'OR' gates.

[c]Or more.

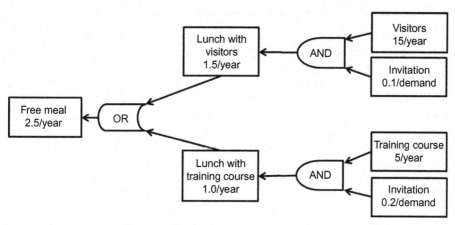

**FIG. 2.4**  Fig. 2.3A redrawn to show different probabilities on different branches.

A common mistake when quantifying fault trees, leading to nonsensical results, is the inappropriate combination of units: frequencies (expressed as events per unit of time) cannot be meaningfully multiplied together (what would be the meaning of '$10^{-3}$ events per "square year"'?), nor can they be added to (dimensionless) probabilities (what would be the result of adding $10^{-3}$ per year and 0.1?). The following are the only allowable combinations of frequency (F) and probability (P):

– OR gate: P+P results in units of P; F+F results in units of F
– AND gate: P$\times$P results in units of P; F$\times$P results in units of F

For example, the combination of F and P at an AND gate typically corresponds to the frequency F of a demand on a safety critical system AND (i.e. multiplied by) the probability P that the system will fail on demand.

Kletz remarks that quantification is generally much less critical than the correct construction of the fault tree.

---

**Text Box 2.24**

… most errors in Hazan are not due to errors in the data but to errors in drawing the fault tree, to a failure to foresee all the hazards or all the ways in which a hazard could arise. Time is usually better spent looking for all the hazards and all the sources of hazard than in quantifying with ever greater precision those we have already found.

In accountancy the figure produced at the end of a calculation, the bottom line, is the one that counts. Risk assessment is different. The way the final figure, the frequency of the top event, is derived is as important, perhaps more important, than the figure itself.

---

Kletz highlighted that arguably the greatest problem with attempting to quantify risk is the accuracy of data available or a lack of information to determine how it can be applied to the scenario being considered.

**Text Box 2.25**

In many Hazans some data are inaccurate, little more than informed guesses, while others are based on a large number of observations. If the inaccurate data have a large effect on the probability of the top event, trying to estimate the others with great accuracy is a waste of time. Yet it is often done. For example, when estimating the effect of a leak of hazardous material, the size assumed for the leak (called the 'source term') is often little more than a guess. Yet very precise and complex calculations are then carried out to find out how it will disperse and what the effects will be. In estimating the probability that an operator will respond to an alarm in the correct way within the required time, people hesitate between estimates of 1 in 10 and 1 in 100, and then use the chosen figure in calculations made to several significant figures.

One of the most valuable exercises that can be carried out when quantifying risks is sensitivity analysis. This allows us to understand the impact of data uncertainty and to focus on the controls that can have the greatest effect on the overall risk. Kletz described this approach.

**Text Box 2.26**

It is important to distinguish between those data that affect the final result significantly and those that do not. Consequence calculations are usually series calculations in which errors in the data are carried through to the final figure. In most probability calculations data from many branches of a fault tree are combined and errors in some data may have little effect.

A very common problem when quantifying risk is incorrectly assuming individual failures are independent. Kletz summarised how dependent (also known as common cause or common mode) failures can occur.

**Text Box 2.27**

Two or three protective systems are never truly independent... For example, all three instruments are from the same manufacturer's batch and have a common manufacturing fault, all three instruments are affected by contaminants in the instrument air or process stream, all three impulse lines are affected by mechanical damage or flooding of a duct, or all three instruments are maintained by the same man who makes the same error. Therefore, we assume that the fractional dead time of a redundant system is never less than $10^{-4}$ (that is, 1 h per year) and is often only $10^{-3}$ (that is, 10 h per year). As we can get $10^{-4}$ with two trips [each with fdt $= 10^{-2}$], a third trip is not worth installing (except as part of a voting system).

For example, wearing a second pair of braces attached to the same buttons may reduce the chance of our trousers falling down. Failure of the buttons (the common mode) is now the biggest cause of failure and adding a third pair of braces, attached to the same buttons, will make no further improvement.

Kletz emphasised the role of diversity in improving safety. Diversity is more effective than redundancy. The more ways that safety systems differ from one another, the less vulnerable they are to common mode failure. Nevertheless, even diversity has its limitations.

---

**Text Box 2.28**

With a diverse system (that is, one in which the approach to a hazardous state is measured in different ways—say by a change in an analysis, a change in pressure and a change in temperature), $10^{-5}$ (6 min per year) may be possible with an extremely complex protective system. For example, belt and braces are better than two pairs of braces. This example illustrates the perils of using thorough mathematics and ignoring practicalities.

Redundancy and diversity are effective when failures are random. They are less effective when failures are due to wear and least effective when failures are systemic.

---

## 2.3.3 Layers of protection analysis

Since the fourth edition of "Hazop and Hazan" appeared in 1999, a new technique for risk assessment has come into widespread use [17–19]. Layers of Protection Analysis (LOPA) is an intermediate form of numerical assessment, filling a gap between simple risk ranking approaches and full QRA. LOPA takes an accident scenario, typically identified in a HAZOP study, and determines whether the basic process controls need to be supplemented by one or more protective systems to address the risk. The IEC61508 and IEC61511 standards cover high integrity protective systems, whose safety instrumented function (SIF) is classified by a safety integrity level (SIL) rating, ranging between SIL-1 (lowest) and SIL-4 (highest). LOPA systematically evaluates the frequency of initiating events for the scenario, the number and nature of independent protection layers (IPLs), the probability of failure on demand (PFD) of each of these layers,[d] and the consequences if all layers should fail, to provide an order of magnitude estimate of the residual risk. This estimate is used to determine if additional layers of protection are required and what degree of integrity is required (i.e. if the residual risk is high, the additional protection needs to have a higher integrity).

The frequency of the initiating event and the PFDs of the IPLs should reflect the conditions on the plant under consideration. Generic values are available as a starting point in various published sources but use of these needs to be properly justified.

It is good practice to avoid the use of SIFs rated above SIL2 and IEC6151 provides guidance on how to achieve this (inherently safer design being the best method).

An important limitation of LOPA is that, unlike FTA, it is incapable of treating dependent (also known as common cause or common mode) failures—although Rothschild [21] has shown how fault tree and LOPA methodologies can be combined into a hybrid analytical tool.

The simplicity of LOPA is deceptive. Practitioners have reported a number of issues, some due to this very simplicity.

---

[d]The use of PFDs implies a low demand rate on the SIFs. High demand and continuous modes (cf brakes on a car) are handled differently [20].

To begin with, authorities disagree whether LOPA should be carried out by a team during or in parallel with a HAZOP, or by a sole expert after the HAZOP, drawing on input from other team members as needed. Whilst the majority appears to favour or at least take for granted the former [22–25], it is noteworthy that Dowell and Bridges, the originators of the technique, have recently set out reasoning in emphatic support of the latter [26]. Kletz might well have agreed with Dowell and Bridges, judging by his contrasting Hazop *"Done by a team"* with Hazan *"Done by one or two people."* Elsewhere, on the other hand, he suggested that *"Hazan is not so esoteric that it can be practised only by an elite band of the initiated… It should be our long-term objective for [engineering] design teams to carry out their own studies."*

There are different approaches to selecting scenarios for LOPA, e.g. every scenario, scenarios with fatality, scenarios that have or require a SIF (to determine the target SIL), and scenarios recommended by the HAZOP team. The last may be most efficient, provided the HAZOP leader thoroughly understands LOPA principles, especially independence, and understands the basics of safety instrumented systems [26].

Where multiple causes lead to a single consequence, common sense demands that their frequencies ought to be added together. However, this is not always straightforward. Care needs to be taken especially in defining the consequence, in order to determine which items should actually be summed. Huberman [27] provides a worked example concerning a forced draft fired heater, where three ways of summing the (mitigated) causes produce three different LOPA results.

Incorrect or even absurd results can be generated by the various abuses of LOPA, most commonly:

- use of generic data on initiating events, without consideration of site-specific factors;
- failure to demonstrate complete independence of protective layers, e.g. the same level device common to more than one protective layer;
- exaggerating or double counting conditional modifiers (see later);
- lack of clarity about risk criteria, i.e. risk of what, from what, and to what;
- confusion about SIL ratings: applying these meaninglessly to individual components rather than to the SIF as a whole [22, 23, 28–30].

There are some concerns that LOPA has oversimplified risk assessment and hence allowed inexperienced personnel to perform calculations [30]. The use of the various spreadsheets that are available for LOPA may be allowing people to plug in data and, as with consequence modelling packages, get answers without fully understanding or thinking through the issues involved. One author [29] writes about how he is appalled at dumbed down pocket 'methodologies' which enable amateurs to establish so-called SIL targets with no need to establish failure mode details or proper quantified risk targets. 'Worse still they are so widely used and taught that it's possible to attend courses in their use and obtain certification – giving the impression of expertise without any proper understanding of the underlying principles and mathematics involved. I frequently encounter 'experts' who can neither explain the difference between a rate and a probability [see discussion in Section 2.3.2] nor establish an appropriate maximum tolerable risk and calculate the maximum PFD required of a risk reduction function'.

A conditional modifier (CM) is defined within LOPA as a feature that can reduce the probability of a hazardous consequence. For example, if the consequence is a Buncefield type vapour cloud explosion following a tank overflow, then one justifiable CM would be the probability of near calm atmospheric conditions at the time of the overflow, since this is now understood to be an essential condition for the formation of a sizeable vapour cloud from this source.

CMs need to be assigned with circumspection. For example, suppose an operator's movements place him close to a potential hazard one-tenth of the time. A CM of 0.1 would seem to be justified if they are only exposed to the risk when in the vicinity and the scenario is not related to the reason they are nearby. But it may grossly underestimate the risk if the scenario is most likely to occur during start-up if the operator is required to be present at that time or, especially, if one of the IPLs is an alarm that requires the operator to carry out a local action, e.g. switching off a pump. The following example [30] shows how a LOPA can be undone by reliance on CMs that appear plausible but are in fact unsound.

The LOPA was concerned with an exothermic batch reaction. If the agitator failed (stopped) whilst reagent was being added, and was then re-started, a runaway reaction was likely. The scenario of most concern was that the operator might notice the agitator had stopped and feel that restarting it was the correct response, without recognising this risk.

In this example, the particular product was only made for approximately 3 months of the year. The LOPA analyst multiplied the initiating event frequency by 0.25 to allow for this. Also, the batch time in the reactor was 14 h but the actual reaction time was 4 h. The analyst therefore relied on another CM of 0.285 (4/14). There was also a bursting disk (sized for the scenario) whose PFD of 0.011 was correctly allowed for. The analyst calculated the event frequency as:

$$0.1\,\text{year}^{-1}\,\left(\text{agitator failure}\right) \times 0.25 \times 0.285 \times 0.01\,(\text{disk}) = 7 \times 10^{-5}\,\text{year}^{-1}$$

At first glance, the use of 0.285 for reaction time to batch time seems reasonable. But what if the agitator failed before reagent addition began? Just hoping that the operator will notice the agitator failure is not reliable and certainly not consistent with the conservative nature of LOPA. If there was an independent speed sensor on the agitator shaft, this would have been used as an IPL in the calculation. Hence, using a CM based on time-at-risk factor of 0.285 is not appropriate.

The factor of 0.25 is also incorrect. If the other processes used in the vessel for the other 9 months were all relatively nonhazardous, the factor of 0.25 would have been a reasonable CM, but the plant in question was multipurpose, hosting a wide variety of reactions, most of which had runaway reaction hazards. So again, this time-at-risk factor is not appropriate and should not have been used.

Ignoring these spurious CMs, the event frequency is $1 \times 10^{-3}\,\text{year}^{-1}$.

Factoring human response into a LOPA can be particularly problematic, including when taking account of response to an alarm when carrying out a SIL assessment. In one case, the assessment team wanted to take credit for the operations team activating the 'big red button' for a compressor following alarm activation. However, when questioned, the control room confirmed that the machine would be closely monitored, but for operational reasons they would wait for the trips to activate rather than shut it down themselves [23].

## 2.4 Consequence analysis

Kletz summarises the main sources of error in Hazan, for each of which he has provided examples or case studies on earlier pages:

---

**Text Box 2.29**

(1) Failure to foresee all the hazards or all the ways in which a hazard can arise.
(2) Errors in the logic.
(3) Failure to foresee that protection may not be fully effective because of poor design or because time of action has been ignored.
(4) Design assumptions not correct; for example, less testing, more demands, failures not random, different mode of operation.
(5) Common mode failures.
(6) Wrong data.

---

Consequence analysis is conspicuously missing from the list. Kletz did, of course, include consequence analysis as an essential step in Hazan, but declined to expand on it in detail, referring the reader to specialist textbooks. He did, however, warn:

---

**Text Box 2.30**

When using these models, it is important to understand the methods they use and their limitations... The biggest uncertainty... is determining the size of the leak. Gas dispersion or explosion overpressure calculations are often carried out with great accuracy although the amount of material leaking out can only be guessed.

---

The second biggest uncertainty would seem to be the vulnerability of the impacted population:

---

**Text Box 2.31**

Is the result of the Hazan in accordance with experience and common sense? If not the Hazan must be wrong. This is obvious, and would not be worth saying if the analysts had not, on a number of occasions, been so carried away by enthusiasm for their calculations that they forgot... to compare them with experience. For example, a number of theoretical studies of chlorine and ammonia releases have forecast large numbers of casualties. When releases have actually occurred, the casualties have been few. Yet the studies do not say this. It was always realised that casualties would be high if conditions were exactly right and this has been tragically demonstrated by the events at Bhopal. However, most toxic gas releases produce nothing like the theoretically possible number of casualties and the reports should state this.

In quantifying the uncertainties associated with each step of consequence assessment of an airborne toxic release, Jirsa [31] confirms Kletz's previous conclusions. Moreover, even in the optimal case, with a robust quality assurance process, removing subjectivity in assessing the characteristics of the release source as much as possible and assuring the proper representation of the range of vulnerability to the toxic exposure within the affected population, and with dispersion only over flat unobstructed terrain, Jirsa finds that the relative error of the complete consequence analysis results will still remain above $\sim$200%.

Jirsa implicitly assumes correct selection and use of consequence models. However, this assumption is often unwarranted. To Kletz's list of error sources in Hazan one might usefully add "(7) Misuse of consequence models".

Dispersion models are widely available to determine the expected concentration of a gas or vapour at varying distances from a leak source. They generally incorporate wind speed as a dilution factor: the concentration downwind from the source of an airborne emission is inversely proportional to the wind speed. Such models therefore break down in calm conditions, since zero wind speed would predict infinite concentration. Some practitioners have attempted to get around this limitation by inputting extremely low wind speeds, e.g. 0.1 m/s. This ignores the random nature of atmospheric interactions, e.g. the fact that, at such low wind speeds, wind direction can vary all over the compass. HSE has sponsored research into dispersion modelling at very low or zero wind speed, conditions that made the Buncefield incident possible [32, 33].

When topography conspires with atmospheric conditions, any off-the-shelf dispersion model is likely to trip up.

The transient venting one night of toxic hydrogen sulphide gas from a reactor on a chemical plant was followed by over a hundred complaints from surrounding residents, many of whom sought medical treatment [34]. Oddly, the complaints arose from all directions, including upwind of the plant. The plant offered to compensate only the minority downwind of the emission and within the path of the plume as assessed using sophisticated dispersion modelling software. However, leaving aside the difficulty of establishing the precise wind direction in the near calm conditions at the time of the release, it was difficult to believe that so many people who were apparently cross-wind or even upwind of the release would be tempted to perjure themselves for the sake of a small compensation. And how did people upwind come to make some of the earliest phone calls logged by the emergency services?

The answer lies in a combination of atmospheric and topographical circumstances: a ground-level inversion and a tall, steep hill near to and downwind of the plant.

On a clear night, the ground cools by radiating heat into space. The lowest layer of the atmosphere is then cooled by contact with the ground. If there is not too much wind stirring the lower atmosphere, this layer will tend to remain at ground level, having become denser through cooling. In turn, it will cool the layer immediately above and thus stabilise it, and so on upwards. Pollutants released into such a stable stratification will be strongly inhibited from dispersing vertically.

The release emerged from a stack 30 m high. However, the summit of the steep hill towards which it was blown by the slight breeze was 65 m high. Calculations showed that the weak airflow carrying the pollutant cloud would not have had enough energy to surmount the hill during a strong ground-level inversion. Consequently, the lower portion of the cloud will have stagnated on the windward side of the hill, gradually spread sideways cross-wind in both directions, cooled—and grew denser—by contact with the ground, and eventually flowed down the hillsides (a phenomenon known as katabatic flow) in a range of directions, including upwind.

Whilst continuing to deny the validity of most of the claims, the plant settled them all before the case came to trial.

Notwithstanding the above, when selected appropriately and used with discretion, i.e. within their validated scope and with due regard for significant complicating factors, computer models for consequence analysis are, of course, an essential element of Hazan. There are some excellent examples of the potential benefits of quantified consequence analysis, including the following three case studies on the design of mitigation systems for hydrofluoric acid (HF) alkylation plant [35]. One case study showed how the use of quantitative consequence data rather than guess work (or engineering judgement) provided a more robust coverage of the detector array to augment existing HF detection, ensuring that enough detectors would be in the path of a potential cloud to satisfy voting logic for automatic activation. Another case study used such data to locate water mitigation equipment and minimise water usage for a site with severe water restrictions. A third study used consequence and process data to segregate process areas in such a way as to more effectively automate acid deinventory.[e]

Recent attempts to address the uncertainties in consequence modelling have shown some promise. One approach [36] involves identifying the sources of uncertainty first and confirming these using sensitivity analysis. Fuzzy logic is then applied to each step of the consequence modelling. The procedure was found to provide less uncertain and more precise results than a deterministic consequence model.

As well as experiments, data from incidents have been used to validate models. In one case, the actual results of 10 accidents that had been thoroughly documented through observed damage to structures and the environment, as well as witness testimony and photographic evidence, were compared with the output of three different consequence models [37]. Although insufficiently accurate for model validation purposes, the data did usefully confirm that the models are reasonably conservative.

## 2.5  Pitfalls in Hazan

Kletz sets out with examples a number of pitfalls, including the use of obsolete or inapplicable generic data, the failure to take into account substandard maintenance or operating

---

[e] The rapid transfer of HF, by the available system pressure or by pumping, from the source vessel to an empty storage vessel.

policy, nonrandom failures (especially in mechanical equipment) and, most insidiously, neglecting to account for dependent failures.

The last of these yields Kletz's "impossibly low fractional dead times" and miraculously low frequencies for the top event, not uncommonly down to the order of $10^{-10}$ per year—or approximately one event since the Big Bang. As Kletz emphasises above, protective systems are never truly independent; however, the modality and degree of dependency is often difficult to assess. Kletz gives us some relatively clear examples:

---

### Text Box 2.32

Watch out for phoney redundancy—parallel or series systems that look as if they're duplicated but the duplication is ineffective.

- Two bursting discs were installed in series so that the failure of one (below the intended failure pressure) would not interrupt production. The upstream one was accidentally installed upside down and it ruptured at a low pressure. The second disc was then ruptured by the shock wave and pieces of the first disc.
- The casing of the Challenger space shuttle was made in two parts, with an O-ring seal between the two parts. Realising that the O-rings were weak features, the designers decided to duplicate them. However, this was ineffective as one ring in a pair is liable to be gripped more tightly than the other.
- If two devices, connected in series or parallel, are tested as a pair then failure is not detected until both have failed. For example, if there are two valves in series and we wish to check that they are isolating, we should check them individually. If we check them as a pair we are not getting the full advantage of redundancy. Two valves in parallel can, of course, be tested as a pair if we wish to check that both are isolating, but not if we wish to check that neither is blocked. Several incidents have occurred on US nuclear power stations because duplicate systems were tested as a unit.

---

One method for addressing less obvious lack of independence of redundant protective systems is to treat them as independent but link the resulting probability of failure via an OR gate to (i.e. add to this probability) a so-called beta factor, representing the estimated dependency [38]. The beta factor may be estimated by expert judgement, either directly or with the aid of checklists developed for the purpose, allowing scoring for the influence of design, operation, and environment. The value of the beta factor is normally determined conservatively, in the range 0.01–0.1.

Finally, even the most detailed and thorough Hazan is only as good as its underlying assumptions (Fig. 2.5).

---

### Text Box 2.33

Few accidents occur because the unlikely odds of one in so many thousand years actually come off. More often, after an accident has occurred, it is found that some of the assumptions on which the analysis was based are incorrect. For example, testing of protective equipment has lapsed or is not thorough, or the faults found are not promptly rectified.

---

**FIG. 2.5**   Unlikely events can still occur.

### 2.5.1 Nonaccidental events

Kletz talks to us from a more innocent age.

---

**Text Box 2.34**

Violations would be better called noncompliances as many (and perhaps most) of them are due to a genuine belief that the rules are unnecessary or inappropriate and that there is a better method of doing the job.

---

Clearly, any intervention by rogue players makes a mockery of Hazan, just as much as incompetent management.

Taking aviation as an example, in 2015 the co-pilot of Germanwings Flight 9525 intentionally flew into the side of a mountain in the French Alps, killing all 150 people on board. He had concealed his mental health crisis from his employer.

In 2018 and 2019 two Boeing 737 MAX aircraft crashed killing 346 people in total. The same technical failure has been cited as the direct cause of both accidents. The evidence suggests that the root cause was a reckless disregard for safety at management level.

The 1999 second edition of Hazop and Hazan has only one index entry relevant to plant security, and that refers to the dubious suggestion[f] that sabotage contributed to the Bhopal disaster (see Section 13 of Appendix). Security concerns are much more prominent now, for example:

- IEC 61511 mandates a security risk assessment on safety instrumented systems and associated equipment, addressing potential threats to security, including cyber security, during the different phases of the design, operation, and maintenance lifecycles [39].
- Rettie et al. have published methodologies for combining studies on the impact of munitions on oil and gas plant with process safety techniques, including event tree analysis, to produce scenario-based risk assessments [40].
- Game theory has been used to bolster security risk assessments of oil and gas pipelines. Social, economic, and political factors are assessed in relation to pipeline 'attractiveness' and vulnerability, enabling the efficient allocation of limited security resources [41].

# References

[1] T. Kletz, Hazop and Hazan—Identifying and Assessing Process Industry Hazards, fourth ed., IChemE, Rugby, 1999.
[2] S. Gadd, D. Keeley, H. Balmforth, Good Practice and Pitfalls in Risk Assessment, HSE RR151, (2003).
[3] Z. Gyenes, M.H. Wood, Lessons learned from major accidents having significant impact on the environment, in: Hazards 24, Paper 17, IChemE Symp Ser 159, 2014.
[4] I. Vince, Major Accidents to the Environment, Elsevier, Oxford, 2008.
[5] G. Atkinson, B. Briggs, Assessment of toxic risks from warehouse fires, in: Hazards 29, Paper 5, IChemE Symp Ser No.166, 2019.
[6] CSB, The report of the BP US refineries independent safety review panel, (2007).
[7] HSE, Evidence or otherwise of scale aversion: public reactions to major disasters. Technical note 03, (2009).
[8] D. Carter, The scaled risk integral—a simple numerical representation of case societal risk for land use planning in the vicinity of major accident hazards, in: Proc. 8th Int. Symp. Loss Prevention and Safety Promotion in the Process Industries, Antwerp, vol. 2, 1995, pp. 219–224.
[9] HSE, Reducing Risks, Protecting People—HSE's Decision-Making Process, (2001).
[10] D.J. Ball, P.J. Floyd, Societal risks—a report prepared for the Health and Safety Executive, (1998).
[11] I. Vince, Societal risk in land use planning—the scale of 'scale aversion', in: Hazards 22, Paper 60, IChemE Symp Ser No.156, 2011.
[12] C.M. Bone, P.H. Bottelberghs, L. Fryer, M.C.M. Hobbelen, I. Vince, G. Pota, et al., Environmental Risk Criteria for Accidents: A Discussion Document for CEC, AEA/CS/16419000/Z/3.1, AEA Technology, 1995.
[13] DETR, Guidance on the Interpretation of Major Accident to the Environment for the Purposes of the COMAH Regulations, (1999).
[14] CDOIF, Guideline—Environmental Risk Tolerability for COMAH Establishments (Version 2.0), (2016).
[15] HSE, Buncefield: Why Did It Happen? The underlying causes of the explosion and fire at the Buncefield oil storage depot, Hemel Hempstead, Hertfordshire on 11 December 2005(2011).
[16] C. Ericson, Fault tree analysis—a history, in: Proceedings of the 17th International System Safety Conference, 1999.

[f] Advanced by consultants on behalf of the plant.

[17] CCPS, Layer of Protection Analysis: Simplified Process Risk Assessment, Center for Chemical Process Safety, 2001.

[18] CCPS, Guidelines for Enabling Conditions and Conditional Modifiers in Layer of Protection Analysis, Center for Chemical Process Safety, 2013.

[19] CCPS, Guidelines for Initiating Events and Independent Protection Layers in Layer of Protection Analysis, Center for Chemical Process Safety, 2015.

[20] H. Jin, W.L. Mostia Jr., A. Summers, High/continuous demand hazardous scenarios in LOPA, in: 12th Global Congress on Process Safety, Houston, TX, AIChE, April, 2016.

[21] M. Rothschild, Fault tree and layer of protection hybrid risk analysis, Loss Prev. Bull. 182 (2005) 18–23.

[22] C. de Salis, Inclusive and integrated risk assessment, risk management and SIF definition under the IEC61508 group of standards, in: Hazards 24, Paper 46, IChemE Symp Ser No. 159, 2014.

[23] J. Fearnley, Where are your SIL assessments now? in: Hazards 24, Paper 48, IChemE Symp Ser No. 159, 2014.

[24] R. Gowland, How to LOPA. Layers of protection—know your onions, Chem. Eng. 899 (2016) 49–53.

[25] A.G. King, LOPA: friend or foe? in: Hazards 25, Paper 67, IChemE Symp Ser No. 160, 2015.

[26] A.M. Dowell, W.G. Bridges, LOPA: performed when and by whom, in: 14th Global Congress on Process Safety, Orlando, FL, AIChE, April, 2018.

[27] A. Huberman, Risk criteria selection and the impacts on lopa results: to sum or not to sum, that is the question, in: 12th Global Congress on Process Safety, Houston, TX, AIChE, April, 2016.

[28] C. Chambers, J. Wilday, S. Turner, A review of Layers of Protection Analysis (LOPA) Analyses of Overfill of Fuel Storage Tanks HSE RR716, (2009).

[29] D.J. Smith, IEC 61508: uses and abuses, TCE 848, (2012) pp. 28–29.

[30] R. Casey, Limitations and misuse of LOPA, Loss Prev. Bull. 265 (2019) 13–16.

[31] P. Jirsa, An analysis of the cumulative uncertainty associated with a quantitative consequence assessment of a major accident, Process Saf. Environ. Prot. 85 (2007) 256–259.

[32] I.G. Lines, D.M. Deaves, Consideration of the Feasibility of Developing a Simple Methodology to Assess Dispersion in Low/Zero Windspeeds, HSE CRR199, (1998).

[33] S. Gant, G. Atkinson, Buncefield investigation: dispersion of the vapour cloud, HSE RR1129, (2018).

[34] I. Vince, T. Fishwick, Beware: the witness may be telling the truth!, Loss Prev. Bull. 264 (2018) 6–8.

[35] J.N. Shah, Unlock the hidden value of QRAs to optimise risk management, in: Hazards 26, Paper 41, IChemE Symp Ser No.161, 2016.

[36] A. Markowski, D. Siuta, Fuzzy logic approach to calculation of thermal hazard distances in process industries, Process Saf. Environ. Prot. 92 (2014) 338–345.

[37] K. Osman, B. Geniaut, N. Herchin, V. Blanchetiere, A review of damages observed after catastrophic events experienced in the mid-stream gas industry compared to consequences modelling tools, in: Hazards 25, Paper 11, IChemE Symp Ser No.160, 2015.

[38] K.N. Fleming, A reliability model for common mode failure in redundant safety systems, in: Proceedings of the Sixth Annual Pittsburgh Conference on Modeling and Simulation, April 23–25, 1975 (General Atomic Report GA-A13284).

[39] IEC 61511-1 edition 2.0 2016-02 Functional safety—Safety Instrumented Systems (SIS) for the process industry sector—part 1: framework, definitions, system, hardware and application programming requirements, (2016).

[40] C. Rettie, M. Vickers, D. McKechnie, Consequence assessment for weapon impacts on process plant, in: Hazards 25, Paper 10, IChemE Symp Ser 160, 2015.

[41] A. Rezazadeh, L. Talarico, G. Reniers, V. Cozzani, L. Zhang, Applying game theory for securing oil and gas pipelines against terrorism, Reliab. Eng. Syst. Saf. 191 (2019) 1–19.

# Inherent safety

## 3.1 Introduction

What you don't have, can't leak

This is the title of a paper written by Kletz in 1978 [1]. It was pivotal in the development of inherent safety. Although Kletz says that he did not develop the original idea it is very clear that he did a great deal to advance and promote it to the point today where it is widely accepted as a fundamental safety concept.

Kletz was such a prolific author that he continued to develop his thoughts and examples on inherent safety in many publications. One aim with this chapter has been to draw together many of these into one place, including Kletz's thoughts on how the absence of inherent safety led to major incidents in the process industries including Flixborough and Bhopal, and other industries including the space shuttle Challenger disaster.

Prompted by Kletz, many other people and organisations have worked on understanding what inherent safety is and how it can be applied. Another aim of this chapter has been to summarise current thoughts on the subject with particular emphasis on practical application.

Inherent safety is a design concept that is best addressed by multifunctional teams in a rigorous and systematic way. This is particularly important where systems or decisions are complex or cross the boundaries of responsibility of different discipline engineers. It is most effective when applied at the earliest stage of any project. Applying inherent safety concepts to an existing system is more difficult and often impractical, but it can still be considered. For example, when specifying or evaluating safety systems it is very useful to recognise the extent to which the design has achieved inherent safety and hence the gap that needs to be filled with the system. Also, when planning work on a system (e.g. maintenance) the principles of inherent safety can be used to define the preparation and safest methods to use.

### 3.1.1 History of inherent safety

A good summary of the history of inherently safer design is provided in the book Kletz wrote with Paul Amyotte on the subject [2]. It suggests that examples can be observed over many years, long before it was recognised as a concept. Kletz provided the following example [2].

---

**Text Box 3.1**

One example of inherently safer design is from the 1870s. In the early days of the Solvay process for the manufacture of sodium carbonate, a manhole at the top of each batch distillation column had to be opened and solid lime tipped in every time the column was charged. Because ammonia vapour could escape, this operation was hazardous. Ludwig Mond suggested that milk of lime should be pumped into the columns instead. However, it was not until 100 years later that inherently safer design was recognised as a specific concept and engineers began to consider it systematically.

---

Kletz described how he first developed ideas of inherent safety in the early 1970s [3].

---

**Text Box 3.2**

One of the papers presented at the conference was a now classical paper by N.A.R. Bell on the manufacture of nitro-glycerine (Loss Prevent in the Process Industries, Symposium Series No 34, Institution of Chemical Engineers, Rugby, 1971, P50). In the old method (illustrated below) acid and glycerine were mixed in a large stirred reactor holding about a ton of raw materials and product watched over by an operator on a one-legged stool. If the temperature got too hot the reactor exploded, taking the operator with it. The reactor was so large because the reaction was slow. Once the molecules came together they reacted quickly. The reaction appeared to be slow because the mixing was poor. Once this was realised it became possible to design a small, well-mixed reactor with the same output as the old one but containing only about a kilogram. The new reactor resembled a laboratory water pump. A stream of acid sucked in the glycerine through a side arm and by the time the mixture left the reactor the reaction was complete. The residence time fell from an hour to a couple of minutes. The maximum size of an explosion was greatly reduced, not by adding on protective equipment that might fail or be neglected but by the inherent or intrinsic nature of the process: what you don't have, can't explode. I was later to quote this paper as an excellent example of inherently safer design—that is, design that are safe because of their nature rather than made safe by adding on layers of protection.

---

Kletz referred to the paper by Bell on many occasions and illustrated how philosophies have changed with the very famous photo of a plant operator sitting on a one legged stool, peering into a reactor. The purpose of the stool had been that if the operator fell asleep he would start to fall, which would wake him up. This was necessary because it was recognised that monitoring the reaction taking place was essential to avoid accidents.

But Kletz said he did not really formulate his thoughts on inherent safety immediately. It took him a few years from reading the paper by Bell for him to recognise its potential and relevance [2].

## Text Box 3.3

A few times in a lifetime, like a single stone setting off a landslide, a single remark makes one look at a whole range of problems in an entirely different way. Kantyka's remark had this effect on me, though the seed he sowed did not germinate until the explosion at Flixborough 4 years later. There were many lessons to be learned from this explosion, but one of the most important was missed by the official inquiry and by most commentators: The leak was big (about 50 tons) and the explosion devastating because there was so much flammable material in the plant (about 400 tons). If the inventory could be reduced, the plant would be safer: What you don't have, can't leak. The inventory was so large because the conversion was low, about 6% per pass, and so most of the raw material, cyclohexane, had to be recovered and recycled; 94% of it got a 'free ride' through the plant—in fact many 'free rides'. If the conversion could be increased, the inventory would be lower.

After this Kletz published his "What you don't have, can't leak" paper [1], which represented a fundamental shift in hazard management towards building safety into the fundamental design of process plants as opposed to adding protective systems to control hazards.

## 3.1.2 Definitions

It is probably fair to say that inherent safety is a concept rather than a clearly defined method or approach. This may explain why the development of a universally agreed definition has not been straight forward.

### 3.1.2.1 *Kletz's definition*

Whilst Kletz wrote a lot about the subject, he does not appear to have used a specific definition of inherent safety. One of the closest attempts appears in his autobiography [3] where he says the main concept is that "it is better to remove a hazard than to keep it under control."

A potentially confusing aspect is that Kletz tended to use the term 'user-friendly design' almost synonymously with 'inherently safer design'. This appears to be a recognition that total elimination of hazard is not possible and so a range of options should be considered to achieve the best overall effect.

### 3.1.2.2 *Definition from CCPS*

The US Center for Chemical Process Safety (CCPS) spent several years developing and refining its definitions of inherent safety. The journey taken is an interesting one because it gets to the heart of what inherently safer design is about.

- 2007 [4]—"Inherently safer - a condition in which the hazards associated with the materials and operations used in the process have been reduced or eliminated, and this reduction or elimination is permanent and inseparable from the process."
- 2009 [5]—"Inherent safety is a concept, an approach to safety that focuses on eliminating or reducing the hazards associated with a set of conditions. A chemical manufacturing

process is inherently safer if it reduces or eliminates the hazards associated with materials and operations used in the process and this reduction or elimination is permanent and inseparable."

- 2010 [6]—"Inherently Safer Technology (IST), also known as Inherently Safer Design (ISD), permanently eliminates or reduces hazards to avoid or reduce the consequences of incidents. IST is a philosophy, applied to the design and operation lifecycle, including manufacture, transport, storage, use, and disposal. IST is an iterative process that considers such options, including eliminating a hazard, reducing a hazard, substituting a less hazardous material, using less hazardous process conditions, and designing a process to reduce the potential for, or consequences of, human error, equipment failure, or intentional harm. Overall safe design and operation options cover a spectrum from inherent through passive, active and procedural risk management strategies. There is no clear boundary between IST and other strategies."

The 2007 definition [4] is attractively simple. A key feature is that the reduction or elimination of the hazard is an intrinsic part of the process and not an added layer of protection. This is probably the most critical feature of the concept and lies at the heart of inherently safer design.

The longer and more complex 2009 definition [5] did not really add much. However, in the same publication [5] it was highlighted that inherent safety is separate from other categories of risk reduction, which may involve passive, active, and procedural controls. This was a useful clarification.

The 2010 definition [6] was clearly more comprehensive than others. Key features include:

- Inherent safety is within a spectrum of risk reduction strategies, which also includes passive, active, and procedural controls,
- There is no clear boundary between inherently safe approaches and the others but passive, active, and procedural controls are different and separate.
- Inherent safety is permanent and inseparable from the process, whereas passive, active, and procedural controls are additional control measures;
- Inherent safety should not be considered in a vacuum but instead balanced with other decision-making criteria, especially where there is significant cost or technical risk.

The final point earlier is also mirrored in UK regulations by the concept of ALARP—As Low As Reasonably Practicable, which is covered later in this chapter.

### 3.1.2.3 *Definition from the UK Health and Safety Executive*

Through his association with Loughborough University, Kletz co-authored a report for the UK Health and Safety Executive (HSE) on the subject of improving inherent safety [7]. This included the following definition:

---

### Text Box 3.4

An 'inherently safe' approach to hazard management is one that tries to avoid or eliminate hazards, or reduce their magnitude, severity or likelihood of occurrence, by careful attention to the fundamental design and layout. Less reliance is placed on 'add-on' engineered safety systems and features, and procedural controls, which can and do fail.

---

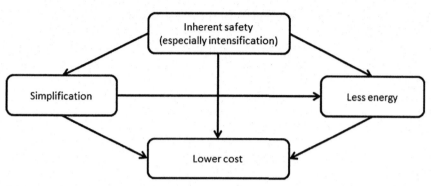

FIG. 3.1    Inherent safety is not a single change but a package of improvements.

### 3.1.2.4 *Definition from the Energy Institute*

In its guidance on applying inherently safe design (ISD) [8] the UK's Energy Institute included the following in its foreword.

> Traditional process safety approaches have often required 'add-on' risk safety systems that are costly to install and maintain. By comparison, ISD provides the opportunity to eliminate hazards or reduce their severity or likelihood by better design, with the potential of reducing overall capital expenditure (CAPEX) and operating expenditure (OPEX). ISD should promote a culture of challenging the need for designs that rely on 'add-on' safety systems, by confirming why they are needed, and how the need could be avoided by improving the basic process design.

## 3.1.3  The benefits of inherent safety

Kletz was very clear on what he saw to be the benefits of inherent safety [9] (Fig. 3.1).

### Text Box 3.5

How much money would we save it we did not have to contain such large inventories or could use safer materials? No one knows. No one keeps separate costs for safety. It is not something added afterwards like a coat of paint, but it is an integral part of the design. If a design engineer is asked the cost of the safety features in his plant he tends to list the newer features which were subject to debate. But he does not include traditional safety features such as relief valves, or even the cost of all the alarms and trips. These features he takes for granted.

As a guess, perhaps 5% of the capital cost could be saved if we could reduce our expenditure on added-on protective equipment such trips, alarms, emergency valves, leak and fire detectors, insulation, water drenching and fire-fighting equipment.

More important would be reductions in the cost of maintaining the equipment and systems. A lot of protective instrumentation means a lot of skilled manpower has to be employed for maintenance and testing. A lot of management time has to be taken up or should be taken up—by monitoring to make sure that equipment is operated and maintained to the necessary standard. Obviously this effort cannot be eliminated completely. But if plants can be made inherently safer, then the effort can be reduced.

Also, it will be easier to persuade the central and local authorities that a plant will not blow up or poison the neighbourhood. It will be easier to find a location for the plant and the equipment may not have to be spread so far apart.

If inventories can be reduced, probably the biggest saving, will not come from a reduction in the amount of protective equipment needed or the effort required to maintain it. Rather, it will come from a reduction in the size of the plant items together with the consequent reduction in structures to support them and land to accommodate them.

## Text Box 3.6

When I worked in industry, my research colleagues sometimes complained about the high cost of extrinsic safety features. They suggested that by recommending so much 'added-on' safety, I was making their processes uneconomic. My reply was that 'added-on' safety was necessary because they invented such poor processes—poor in other ways beside safety. It is not economic to push large inventories round and round. If they could invent better processes we would not need to add so much safety equipment.

## 3.2  The concept of inherent safety

Whilst there is consistency between the various definitions listed earlier, do they convey the true principles of inherent safety? In many ways it is easier to understand it by having a better grasp of the aims and objectives.

Kletz summed up his views of the concept of inherent safety as follows [10].

## Text Box 3.7

The essence of the inherently safer approach to plant design is the avoidance of hazards rather than their control by added protective equipment. Although we generally think of safety in a comparative sense, one experienced practitioner[a] has made the distinction between safe design and safer design. He explains that with safe design, there are active safeguards to prevent the occurrence of hazardous events and to protect people and plant from the effects. With safer design, there are fewer hazards, fewer causes, and fewer people to be exposed to the effects.

[a]G. Dalzell, Inherently safety design: changing attitudes and relationships. Seventh International Conference on Health, Safety and Environment in Oil and Gas Exploration and Production. Canada Society of Petroleum Engineers. 2004.

But in true Kletz style he illustrated this concept with some very simple but effective statements:

1. "What you don't have, can't leak"
2. "People who are not there can't be killed"
3. "The more complicated a system becomes, the more opportunities there are for equipment failure and human error"

Kletz's explanation of these concepts and how he relates them to industrial accidents helps us to understand the thoughts that drove the definitions used commonly in industry.

## 3.2.1 What you don't have, can't leak

This statement represents a very significant step in both Kletz's and industry's understanding of inherent safety and why it is so important. He explained what it means in practice [11].

### Text Box 3.8

The best way of preventing a leak of hazardous material is to use so little that it does not matter if it all leaks out, or to use a safer material instead. We cannot always find ways of doing this but once we start looking for them we find a surprisingly large number.

Until the explosion at Flixborough, UK in 1974, which killed 28 employees the attitude in the process industries was, 'There is no need to worry about large stocks of hazardous materials as we know how to keep them under control'. Flixborough weakened that confidence and 10 years later the toxic gas release at Bhopal, India, which killed over 2000 members of the public, almost destroyed it. Since then companies, to varying extents and with varying degrees of success, have tried to reduce their stocks of hazardous materials in storage and process.

To do so requires a major change in the design process: much more consideration of alternative in the early stages of design, and this will not occur without encouragement and support from the senior levels of management.

#### 3.2.1.1 *Flixborough accident*

On 1 June 1974 the Flixborough chemical plant in North Lincolnshire, UK exploded killing 28 people and seriously injuring 36. The plant site produced caprolactam, a chemical used in the manufacture of nylon. The explosion occurred due to pipework failure that led to a release of highly flammable cyclohexane.

The normal focus on this accident is the failure to manage plant modifications but Kletz also used it to discuss the role of inherent safety [10].

### Text Box 3.9

The explosion at Flixborough in 1974 occurred in a plant for the oxidation of cyclohexane with air, at about 150°C and a gauge pressure of about 10 bar (150 psi), to a mixture of cyclohexanone and cyclohexanol, usually known as KA (ketone/alcohol) mixture. It is a stage in the manufacture of nylon. The inventory in the plant was large (200–500 tonne has been quoted) because the reaction was slow and the conversion low, the latter being about 6% per pass! Much of the inventory was held in six large continuous reactors operated in series, and the rest was held in the equipment for recovering the product and recycling the unconverted raw material.

The first stage of the reaction (hydroperoxide formation) was slow because mixing was poor. Conversion was kept low for the same reason. If the oxygen concentration in the liquid is high it can easily become too high where it leaves the gas sparger, and then unwanted side reactions (further oxidation of the cyclohexane) occur. A method of improving the mixing in gas/liquid reactors has been described. In most such reactors the gas is added through a sparger, and the liquid is stirred by a conventional stirrer. In the Litz design the gas is added to the vapour space, and a down-pumping impellor sucks the gas down and mixes it intimately with the liquid. Unreacted gas that escapes back into the vapour space is recirculated. The reactor was designed to increase output and efficiency, but compared with a conventional reactor less inventory is needed for a given output. Other methods of improving mixing are suggested. The second stage of the reaction, decomposition of the hydroperoxide, is inherently slow, and a long residence time is required. Could this be achieved in a tubular reactor as discussed above? Higher temperatures would increase the rate of decomposition.

Kletz went on to say that the inherently safe solution is not always obvious and can raise other issues that need to be considered. For example, addressing the issues with the second stage reaction would have required a new approach to heat removal. But if this had resulted in a significantly lower hazardous inventory the impact of any accident would have been greatly reduced.

Kletz also used Flixborough to highlight that there can be unintended consequences of changes, with transfer of risk being a particular concern [9].

### Text Box 3.10

At one extreme, the 'back to nature' enthusiasts suggest we abandon man-made fibres and go back to the use of wood, wool and cotton. They overlook the fact that the 'accident content' of natural materials is higher than that of synthetic materials, though people are killed one at a time. (The price of any article is the price of the labour used to make it, capital costs being other people's labour. Agriculture is a low wage industry so there will be more hours of labour in a wool or cotton shirt than in a synthetic fibre shirt of the same price. Because agriculture is also a high accident industry this means that there will be more fatal accidents per woollen or cotton shift than per nylon shirt.)

Given that the demand for nylon would remain, the occurrence of the accident prompted a review of how it could be made more safely. Kletz shows how the chosen solution probably raised as many questions as it answered [9].

### Text Box 3.11

An example often quoted of an inherently safer route is the manufacture of a mixture of cyclohexanone and cyclohexanol, usually known as KA or ketone-aldehyde mixture. It is an intermediate in the manufacture of nylon. At Flixborough it was manufactured by the air oxidation of cyclohexane. A leak from the plant, the result of a substandard modification, exploded, destroying the plant and killing

28 people. When the plant was rebuilt cyclohexanol was manufactured by an alternative route, the hydrogenation of phenol. This is a vapour phase process and a much less hazardous one than the oxidation route KA. However, the phenol has to be manufactured, usually by the oxidation of cumene to cumene hydroperoxide and its 'cleavage' to phenol and acetone. This process is at least as hazardous as, perhaps more hazardous than, the oxidation of cyclohexane and as much extrinsic safety equipment is required. It was not carried out at Flixborough but elsewhere. There was less hazard on the Flixborough site, but no reduction in the total hazard of the process.

### 3.2.1.2 Bhopal

Kletz was very clear on his views on the Bhopal disaster and how even a basic understanding of inherent safety would have made a great difference to the consequences [12].

### Text Box 3.12

Methyl isocyanate, the material that leaked and killed over 2000 people, was not a raw material or product but an intermediate. It was convenient to store it but not essential to do so. If it had been made continuously and used as it was made, the worst possible leak would have been a few kilograms from a ruptured pipeline. After Bhopal many companies did reduce their stocks of hazardous intermediates. Alternatively, the production of methyl isocyanate could have been avoided by reacting the three raw materials in a different order.

### 3.2.1.3 Ammonia as a refrigerant

Ammonia is commonly used as a refrigerant, but it is a severe irritant and human exposure can be fatal. In one accident 15 people were killed when liquid ammonia leaked from a refrigeration plant in Shanghai [13]. The leak was caused when a cap on a pipeline carrying ammonia was detached. Although the investigation concluded that the company concerned had not conformed to safety requirements it was clear that use of another, less hazardous refrigerant would have had less severe consequences.

Unfortunately alternative refrigerants such as Chlorofluorocarbons (CFCs) and Hydrochlorofluorocarbons (HCFCs), that were used in domestic refrigerators because they are safer to people, were subsequently found to be very hazardous to the environment and have been banned internationally. This highlights the challenges of inherent safety. In this case there is a requirement to weigh up safety and environmental risks [13].

## 3.2.2 People who are not there can't be killed

Whilst hazard elimination will always be the most effective measure, Kletz was very well aware that this was not always possible or desirable. With this in mind he pointed out that keeping people away from hazardous areas can be very effective at reducing the consequences of any accidents that occur [11].

**Text Box 3.13**

Just as material which is not there cannot leak, people who are not there cannot be killed. Those killed at Bhopal were living in a shanty town which had grown up close to the plant. Earlier the same year 542 people were killed in a BLEVE in Mexico City. Most were living in a shanty town near the plant.

It is difficult in some countries to control development. Nevertheless plants handling large amounts of hazardous materials should not be built unless concentrations of people can be kept well away. Fortunately disasters such as Bhopal can occur in only a few plants and industries but the principle of segregation applies everywhere. By sensible lay-out—for example, by keeping people and traffic apart or by not putting a workshop near a unit that might explode—we can reduce the chance that someone will be killed or injured.

Using Bhopal and Mexico City accidents as examples of keeping people outside of the hazardous area implies that this is only a concern for people living, working, and visiting places near to, but external to the site. In countries with effective arrangements for land use planning this is handled by regulations. Unfortunately this does not apply in all countries and companies can have little control over who could be exposed to the hazards they handle. However, companies do have control over who enters their site and where they work and visit. Understanding the hazardous areas and rearranging the site and working arrangements can be effective at reducing the potential consequences of an accident.

### 3.2.2.1 Piper Alpha

The fires and explosions at the Piper Alpha oil platform in July 1988 led to the deaths of 167 people partly because there was a single structure that housed equipment used for processing hydrocarbons, a range of work spaces, and the accommodation for the people working there. Although some protection had been provided in the form of fire walls, these proved to be ineffective due to the scale of the incident.

Multijacketed (a 'jacket' is the structure which supports an offshore platform) facilities have become far more prevalent in the offshore oil and gas industry. These provide different structures to separate accommodation including offices from the hydrocarbon. They do not eliminate the exposure to the hazard because people still have to visit the other jacket to do their work but if there were to be an incident there would be fewer people present and they have a permanent means of escape to a safer location.

### 3.2.2.2 BP Texas City refinery accident

Most of the 15 people killed as a result of the explosion and fire at the Texas City refinery in 2005 were located in a temporary building (referred to as a 'trailer' in the United States). The plant was being started up at the time of the accident, which is known to be a vulnerable time due to process instability as fluids are introduced and pressures, temperatures, and flows are increased. Due to the increased risk it would have been prudent to have removed all personnel from the buildings whilst the start-up was taking place. However, the inherently safer solution would have been to locate the building at a safe distance from the plant. In this case it may have not been considered necessary because the building was only temporary, but this accident clearly highlights the danger of this approach.

### 3.2.2.3 *Corden pharmachem fatal accident*

On 28 April 2008 a runaway chemical reaction caused distortion of a pressure rated reactor and led to a loss of containment, which created a blast wave sufficient to kill one person, seriously injure another, and cause significant physical damage [14]. The accident highlighted a number of failures that build the case for inherent safety, including the runaway occurring because a solvent (acetone) was omitted in error and the installed relief valves not being able to cope. However, people were harmed because, when a runaway reaction was detected, it was necessary for operators to enter the plant area to shut manual valves [14]. If remotely operated valves had been available these people could have remained in a safer location and may not have been harmed when the reactor failed.

## 3.2.3 The more complicated a system becomes, the more opportunities there are for equipment failure and human error

This final concept due to Kletz was a recognition that complexity can be a cause of problems [11].

---

**Text Box 3.14**

The usual response to this statement is that complication is inevitable today. Sometimes it may be, but not always; There are many ways in which plants have been made simpler, and thus cheaper and safer. As with the reduction of stocks, the constraints are often procedural rather than technical. We cannot simplify a design if we wait until it is far advanced; we have to consider alternatives in a structured and systematic way during the early stages of design.

---

Kletz highlighted that one of the causes of complexity is the safety systems added to hazardous systems [11].

---

**Text Box 3.15**

Many plants contain a large amount of hazardous materials. We try to keep it under control by adding on trips and alarms and other protective equipment which may fail or can be neglected. It would be better to devise processes and equipment that use less hazardous material, so that it does not matter if it all leaks out, or safer materials instead.

---

A quote that sums up this concern about complexity was made in relation to software design [15]. It is interesting to note that this was said in 1980, when by today's standards the computer technology was so much simpler:

There are two ways of constructing a software design: One way is to make it so simple that there are obviously no deficiencies, and the other way is to make it so complicated that there are no obvious deficiencies. The first method is far more difficult. It demands the same skill, devotion, insight, and even inspiration as the discovery of the simple physical laws which underlie the complex phenomena of nature.

### 3.2.3.1 *Avoiding static discharges*

On 29 October 2007 a fire and series of explosions occurred at the Barton Solvents chemical distribution facility in Des Moines, Iowa, United States. The investigation concluded that a vapour–air mixture was ignited due to static discharge (spark) occurring between the body of a 'tote' tank being filled with ethyl acetate and a metal component of the equipment being used for filling [16].

Many things can be done to reduce the likelihood of static discharges including ensuring proper electrical bonding and/or grounding of equipment and controlling flow rate. However, splash filling a vessel (as was the case in this accident) means it is always possible. A more inherently safe option is to avoid splash filling by using a dip pipe that can be inserted into the top but will reach to the bottom of the vessel. This will avoid splashing and hence the generation of static [16] (Fig. 3.2).

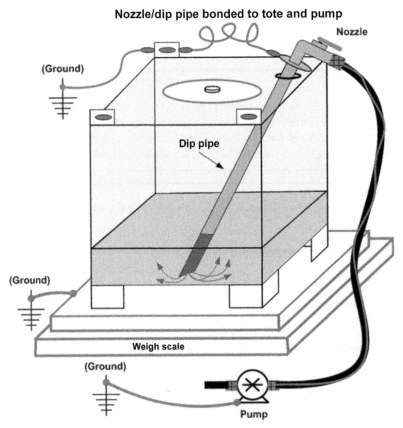

**FIG. 3.2**   Image from CSB report [16] showing how to avoid splash filling and reduce risk of static.

### 3.2.3.2 *Chevron Richmond refinery*

On 6 August 2012 a catastrophic pipe rupture in a crude unit at the Chevron USA Inc. Refinery in Richmond, California, United States resulted in a large fire that engulfed 19 workers. All managed to escape with six experiencing minor injuries [17].

The pipe ruptured due to sulphidic corrosion, which occurs when steel is exposed to sulphur compounds at certain temperature ranges. It is a well-known phenomenon in refineries due to the sulphur content of crude oil and is avoided by using appropriate materials of construction wherever the risk justifies it. When initially designed, the pipe that failed in this accident was not considered to pose a significant risk due to the types of crude expected to be processed and so was constructed from carbon steel. However, operating conditions changed over the life of the refinery and the risk of sulphidic corrosion increased. This had been recognised by some people within Chevron, who recommended increased inspection regimes. This would have been a complex undertaking and was not implemented [17].

In this case implementing an inherently safer design by selecting materials of construction that would be resistant to sulphidic corrosion would have simplified management of changes to process conditions and would not have required any increased inspection regimes above those considered 'normal' in a refinery.

## 3.3 Principles of inherent safety

Whilst the concepts described earlier were effective at illustrating why inherent safety was important they did not give such a clear idea of how it can be achieved in practice. To do this Kletz developed the following set of principles:

- Intensification—lower inventories of hazardous materials;
- Substitution—use a less hazardous substance;
- Attenuation—use less hazardous conditions;
- Limitations of effects—design to minimise or withstand hazardous circumstances so that there is less reliance on additional protective devices;
- Simplicity—less equipment that can fail, and equipment design that is easier to understand and operate as intended (so human error is less likely).

These principles remain evident today although some of the terminology has changed. In particular 'intensification' has been replaced by 'minimisation' and 'attenuation' has been replaced by 'moderation'.

Other principles have been introduced. For example, a list from the Energy Institute [8] includes 'elimination' and 'segregation', which do not appear on Kletz's list.

### 3.3.1 Intensification or minimisation

The aim here is to perform the same activity with smaller quantities of hazardous material or performing an activity less often [8]. Kletz illustrated this principle with the following image [9] (Fig. 3.3).

FIG. 3.3    Intensify to increase inherent safety.

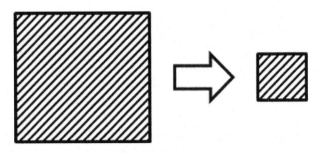

Kletz provided a number of examples of how intensification or minimisation can be applied in practice including [9]:

- Reducing reactor sizes by increasing reaction rates though improved mixing, increasing temperature or pressure, or switching from liquid phase to vapour phase reactions;
- Reducing the size of distillation columns by selecting different types of packing or tray design, reducing the quantity of liquid by reducing the diameter of the column base in relation to higher up the column where vapour will be present, using internal auxiliaries such as reboiler or bottoms pumps, and considering other forms of separation;
- Reducing inventories in heat exchangers by selecting different design (e.g. printed circuit instead of shell and tube);
- Reducing inventories in liquid–liquid separation systems by using devices to promote coalescence; and
- Reducing quantities of materials stored, with the aim of eliminating all unnecessary intermediate storage.

Kletz provided the following comparisons of reactor design that sum up the types of considerations that should be made when looking for an inherently safer solution [9].

---

**Text Box 3.16**

No unit operation offers more scope for reduction of inventory than reaction. Many reactors, particularly liquid phase oxidation reactors, contain large inventories of hazardous materials and leaks from them have caused many fires and explosions, including that at Flixborough.

---

Reactors can be arranged in the following hierarchy of (usually) increasing inventory and decreasing inherent safety (Table 3.1).

---

**Text Box 3.17**

In a batch reactor, all the reactants are added before reaction starts. In a semibatch reactor, one (or more) of the reactants is added and the final reactants added gradually as reaction proceeds; reaction takes place at once, and an unreacted mixture cannot accumulate unless mixing fails or the catalyst is consumed.

**TABLE 3.1** Hierarchy of reactor safety [2].

| | |
|---|---|
| Vapour-phase reactors | |
| Liquid-phase reactors | Inventory decreases |
|    Thin-film reactors | |
|    Tubular reactors (once-through and loop) | |
|    Continuous pot (stirred tank) reactors | Safety increases |
|    Semibatch reactors | |
|    Batch reactors | |

Each process should be considered on its merits, for there are exceptions to the general rules. For example, according to Englund[b], for the copolymerisation of styrene and butadiene, a semibatch reactor is safer than a batch one, but a continuous reactor may not be the safest because some designs contain more unreacted raw material than semibatch reactors.

When possible, vapour-phase reactors should be developed in place of liquid phase ones because the density of the vapour is less and the leak rate through a hole of a given size is lower. Of course, a gas at very high pressure with a density similar to that of a liquid is as hazardous as a liquid.

As a rule, reactors, of all types are large not because a large output is desired, but because conversion is low, reaction is slow, or both.

### 3.3.1.1 Incident caused by a shared water flushing system

Kletz shared the following example of where intensification or minimisation of the water supply to three gasifiers could have prevented an accident [18] (Fig. 3.4).

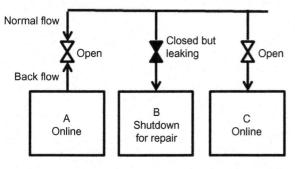

FIG. 3.4 Shared pipework increases vulnerability to isolation failure.

[b]S. Englund, Design and operate plant for inherent safety, Part 1. Chem Eng Progress 87. 1991.

## Text Box 3.18

Another incident involving shared equipment occurred on an ammonia plant on which there were three parallel gasifiers, A, B and C. They were distinct units but a common 1 in. (25 mm) line supplied water at a gauge pressure of 63 bar (910 psi) for flushing out level indicators and level gauges. The plant operated at a gauge pressure of 54 bar (780 psi). Gasifiers A and C were on line and B was shut down for repair. A large amount of flushing water was used on C and the pressure in the water line near A and B fell below 54 bar. Gas from A entered the water line, entered B through a leaking valve and came out inside B where it was ignited by a welding spark. Fortunately no one was injured. The connection from the water line to B gasifier had not been slip-plated as no one thought it possible for gas to enter by this route. Afterwards separate water lines were run to the three gasifiers.

## 3.3.2 Substitution

The aim here is to reduce the hazard severity by replacing a hazardous substance or a processing route with a less hazardous alternative. Another option is to replace a procedure with one that presents a lesser hazard [8] (Fig. 3.5).

Kletz provided a number of examples of how substitution can be applied in practice including [9]:

- Using safer agents that are less flammable, toxic, or reactive in services such as heat transfer or refrigeration, or as a solvent;
- Choosing a process that uses less hazardous materials or less hazardous operating conditions.

Kletz provided the following example of substituting a flammable hydrocarbon heating medium with water as an inherently safe solution [9].

## Text Box 3.19

Flammable hydrocarbons or ethers are often used as heat-transfer media at pressures considerably above atmospheric. In some plants the inventory of the heat transfer system is several hundred tonnes and exceeds that in the process. Sometimes the heat transfer medium is heated in a furnace and supplies heat to a reactor or distillation column reboiler; sometimes it removes heat from a reactor and gives it up to water in a cooler, possibly raising steam.

FIG. 3.5 Substitute to increase inherent safety.

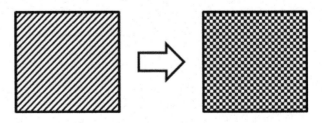

When possible, we should use a higher boiling liquid or, better still, water. Though pressures will be higher than in a hydrocarbon or hydrocarbon/ether system, the technology is well understood and if the heat-transfer medium is used to remove heat from a reactor, the heat is immediately available as steam which can usually be used elsewhere on the site.

### 3.3.2.1 Bhopal

The Bhopal tragedy highlights why inherent safety has to be considered for any hazardous process. Kletz discussed the substitution options that were available that could have prevented or significantly reduced the consequences [18] (Fig. 3.6).

Text Box 3.20

If reducing inventories, or intensification as it is called, is not practicable an alternative is substitution, that is, using a safer material or route. At Bhopal the product (carbaryl) was made from phosgene, methylamme and alpha-naphthol. The first two were reacted together to make MIC which was then reacted with alpha-naphthol. In an alternative process used by the Israeli company Makhteshim, alpha-naphthol and phosgene are reacted together to make a chloroformate ester what is then reacted with methylamine to make carbaryl. The same raw materials are used but MIC is not formed at all.

Of course, phosgene is also a hazardous material and its inventory should be kept as low as possible, or avoided altogether. If carbaryl can only be made via phosgene, perhaps another insecticide should be manufactured instead.

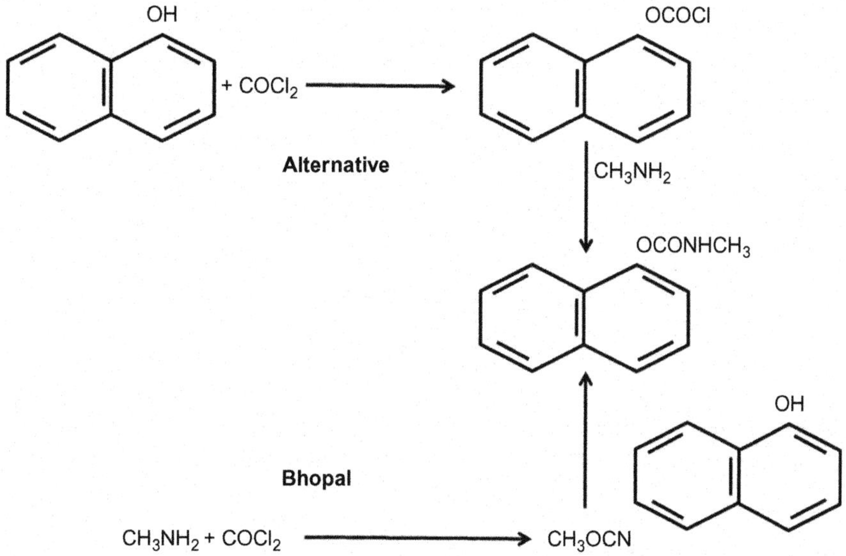

FIG. 3.6   Routes to making carbaryl.

Or instead of manufacturing pesticides perhaps we should achieve our objective—preventing the harm done by pests—in other ways, for example, by breeding pest-resistant plants or by introducing natural predators? I am not saying we should, both these suggestions have disadvantages, merely saying that the question should be asked.

### 3.3.2.2 *Challenger space shuttle*

On 28 January 1986 the Challenger Space Shuttle exploded killing all seven crew members. The direct cause was a failure of rubber O-rings that allowed fuel to escape from the rocket boosters. Kletz used this as an example of where a substitution of components could have prevented the accident and the difficulties of managing additional layers of protection if the most inherently safe option is not selected [18].

---

**Text Box 3.21**

NASA accepted the cheapest quotation for the design of the shuttle even though the casing would be made in parts. One of the four quotations was for a one-piece design, an inherently safer one as its integrity does not depend on additional pieces of equipment, seals, which are inevitably weaker in nature than the parts they join. The failure of one of these O-ring seals led to the loss of Challenger.

Realising that the O-rings were weak features, the designers decided to duplicate them. However, this did not give them the redundancy they thought it would as one ring in a pair may be gripped less tightly than the other one. The first flights showed that the O-rings became eroded but this caused little concern amongst the senior management. Development of a new design, a 27-month programme, was started but meanwhile the old design continued in use, for several reasons.

---

The O-ring option may have been cheaper but relied on additional controls. The quality of the O-rings installed required very close attention. Another factor that was not recognised or accepted was that O-ring reliability was affected by ambient temperatures. If the full cost of the quality control and the operational limits on ambient temperature had been properly accounted for it is likely that the apparently higher costs of the inherently safer solution may actually have been cheaper than the O-rings over the life of the Space Shuttle programme.

There were a number of cultural issues that resulted in Challenger being launched when the ambient temperature was too low, which resulted in the O-ring failure. This raises an interesting observation. Organisations with poor safety culture are least likely to recognise the importance of inherent safety. However, these same organisations have the most to gain in applying inherent safety because their poorer safety culture makes them more prone to operational failures. On the flip side, organisations with a better safety culture gain additional benefits because selecting inherently safer designs gives them more time to focus on managing their residual risks.

### 3.3.2.3 *Substituting materials of construction*

In April 2010 a heat exchanger ruptured at Tesoro's Anacortes refinery in Washington State, United States resulting in a fire and explosion. Seven people died [19]. The rupture occurred due to a reduction in the strength of steel caused by High Temperature Hydrogen Attack (HTHA).

The investigation report [19] stated that "It is very difficult to inspect for HTHA because the damage might not be detected" and that "Successful identification of HTHA is highly dependent on the specific techniques employed and the skill of the inspector, and there are few inspectors who have this expertise. Inspection is therefore not sufficiently reliable to ensure mechanical integrity and prevent HTHA equipment damage." It went on to say that chromium steels are far more resistant to HTHA than carbon steel and that using it would have been an inherently safer solution and hence a much better solution. [19]

## 3.3.3 Attenuation or moderation

The aim here is to use a substance in a way that reduces its hazardous properties or to use less severe processing conditions [8]. Another way is to store or transport material in a less hazardous form. Kletz illustrated this principle with the following image [9] (Fig. 3.7).

Kletz provided a number of examples of attenuation [10]:

- Attenuate reactions by controlling the operating temperature to below where a runaway reaction can occur, replace a batch with a continuous reactor, where the improved mixing reduces the required contact time of the reagents, or select catalysts that allow reactions to take place at less severe operating conditions;
- Attenuated storage and transport can be achieved by transforming a hazardous powder into a paste, dissolving a hazardous material in a solvent, or liquefying via refrigeration instead of pressurisation.

Kletz gave the following explanation for using refrigeration over pressurisation [9].

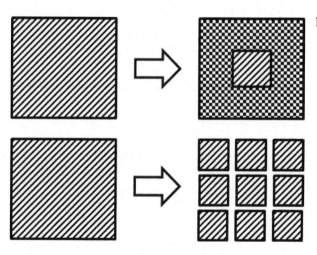

FIG. 3.7   Attenuate to increase inherent safety.

**Text Box 3.22**

Large quantities of liquefied ammonia and chlorine are now usually stored refrigerated at atmospheric pressure, instead of being stored under pressure at atmospheric temperature. If there is a hole in the plant, the quantity discharged is much less; if there is a large spillage, the amount of vapour produced is much less.

### 3.3.4 Limitations of effects

The aim here is to design equipment and processes so that the effects of equipment, control system, or human failures are limited [10]. Kletz gave the following examples:

- Equipment design—spiral-wound gaskets that leak at a lower rate than fibre gaskets, dikes, or bunds having higher walls so that they can have a smaller surface and hence lower evaporation rate of spilt liquid, not overdesigning pipework so that held-up inventories are always minimised;
- Changing reaction conditions—carrying out different stages of a reaction in different vessels if this reduces the potential for incompatible materials to be mixed, ensuring reagents are added to a reactor in a way that minimises the potential for hazardous reactions, changing temperature, concentration, or another parameter;
- Limiting the energy available—using a heating medium that is incapable of overheating the process, and so reducing the risk of fire or runaway reaction, using pumps with discharge pressure less than the receiving vessel's design pressure or relief valve setting;
- Elimination of hazards—using fully flooded drainage systems that do not have a vapour space that can create a flammable atmosphere, using submerged pumps directly inside tanks to remove the need for a pump room, moving all potential ignition sources to outside a flammable hazard area, reducing the number of samples taken to only those necessary, reducing maintenance frequency through better design and planning.

#### 3.3.4.1 *Flixborough and Bhopal*

Kletz again used the Flixborough and Bhopal accidents to highlight how inherent safety could have been applied, in this case by reducing the consequence through limitations of effect [18].

**Text Box 3.23**

It is almost impossible to prevent ignition of a leak the size of that which occurred at Flixborough. (Not all such leaks ignite, but it is often a matter of chance which ones do or do not.) The leak may spread until it reaches a source of ignition. However, it is possible to locate and lay out a plant so that injuries and damage are minimised if an explosion occurs. Most of the men who were killed were in the control room and were killed by the collapse of the building on them.

Is it right to protect the men in the control room and ignore those who are outside? At Flixborough most of those outside survived. Men in an ordinary unreinforced building are at greater risk than those outside.

We now move to the Bhopal incident.

---

**Text Box 3.24**

If materials which are not there cannot leak, people who are not there cannot be killed. The death toll at Bhopal would have been much smaller if unauthorised developments had not been allowed to grow up near the plant. In many countries planning controls prevent such developments, but not apparently in India.

---

It is easy to casually and wrongly think that these problems at Bhopal were because it happened in India and would not have occurred in a wealthier or more developed country. However, Kletz stated [18]:

---

**Text Box 3.25**

In one respect the managing director of Union Carbide, India showed more awareness that his US colleagues: he queried the need for so much MIC storage but was overruled.

---

### 3.3.4.2 *Sterigenics ethylene oxide explosion*

Ethylene oxide is widely used as a sterilising agent but it presents a particularly serious explosion hazard. This has become more of an issue as a result of increasingly stringent requirements for pollution control [20].

On 19 August 2004 and explosion occurred at the Sterigenics International Inc. sterilisation facility in Ontario, California, United States. Four workers were injured and the facility was severely damaged [21]. It occurred when a door to a steriliser was opened prematurely, which admitted air and allowed an explosion mixture with ethylene oxide to form. A computer control system was supposed to prevent this occurrence but it was overridden by maintenance personnel. This highlights how safety systems can always be defeated and so emphasises the value of applying inherent safety wherever possible.

In this case, using a less hazardous sterilisation agent would have been preferred. But this will not always be possible. The potential for explosion should always be a concern due to the flammability of ethylene oxide. Active devices can be used for detecting hazards and initiating an appropriate response, but of course these can fail. Passive devices would be inherently safer. In this case flame arrestors have been proven to be effective at stopping explosions and specially designed valves are available that do not allow explosions to propagate [20].

### 3.3.5 Simplicity or simplification

The aim here is to reduce the likelihood of an accident through inherent features of the design. This can involve designing processes, equipment, and procedures to eliminate opportunities for failure, including human error; also, designing equipment that cannot be exposed to extreme process conditions by the worst-case processing conditions [8]. Kletz illustrated this principle with the following image [9] (Fig. 3.8).

**FIG. 3.8**   Simplify to increase inherent safety.

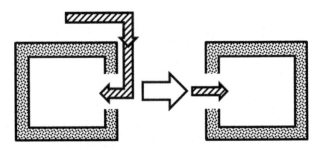

Kletz felt that complexity occurs because inherently safer designs are not adopted and so additional safety devices have to be added. This becomes a particular issue if safety studies are performed late in the design process because the opportunity to adopt an inherently safer design has already been missed. He gave the following examples of simplification [9]:

- Using stronger equipment so that relief devices are not required;
- Using materials resistant to harsh conditions (e.g. extreme temperature, corrosivity) so that reliance on operating procedures and safety systems is reduced;
- Simplifying system arrangements so that the likelihood of human error and vulnerability to instrumentation failures are reduced;
- Relocating potential ignition sources to outside of flammable hazard zones;
- Incorporating passive arrangements to avoid the need for active safety devices.

### 3.3.5.1 Esso Longford

Kletz shared the example of the explosion at Esso's Longford site as an example of where inherent safety using simplification may have been beneficial [18].

---

**Text Box 3.26**

On 25 September 1998 a heat exchanger in the Esso gas plant in Longford, Victoria, Australia fractured, releasing hydrocarbon vapours and liquids. Explosions and a fire followed, killing two employees and injuring eight. Supplies of natural gas were interrupted throughout the State of Victoria and were not fully restored until 14 October. There was no alternative supply of gas and many industrial and domestic users were without fuel for all or part of the time that the plant was shut down.

[Information about this accident] is based on the official Reports and on a book by Andrew Hopkins[c]. The Report is commendably thorough. It describes the operator failings that led to the explosion but does not blame the operators, as reports in earlier time might have done, but shows how the operators' failings were the result of inadequate training and other management failures. In contrast, the weaknesses in design get less consideration.

[c]A. Hopkins. Lessons from Longford—The Esso Gas Plant Explosion. CCH Australia Limited (2001).

As the result of a plant upset the lean oil pump stopped and was not restarted for several hours. There was now no flow of warm lean oil through the heat exchanger and its temperature fell to that of the rich oil, $-48°C$. The Report describes in great detail the circumstances that led to the pump stopping but all pumps are liable to stop from time to time. There are many possible causes such as mechanical failure, loss of power supply, low level in the suction vessel and high level in the delivery vessel. All possible causes should be recognised and precautions taken. The precise reason why the flow stopped on 25 September 1998 is of secondary importance. Next time the pump will stop for a different reason. In this case one of the underlying reasons was the complexity of the plant. It had been designed to recover as much heat as possible and this resulted in complex interactions, difficult to foresee, between different sections. Whatever the reason for the pump stopping, the consequences of the interruption of flow were of major importance.

Equipment is, of course, designed to withstand the temperature (and other conditions) at which it normally operates. Today, it is often designed to withstand the temperatures that it can reach under abnormal or fault conditions. This was less common in 1969 when the plant was built. It was more common then to rely on control and trip systems and operator intervention to prevent equipment reaching temperatures that could cause failure, either immediately or in the long term.

### 3.3.5.2 *Refinery furnace failure*

In May 2013 a major failure occurred during the start-up of a primary reformer in an oil refinery in Southern Africa [22]. Although no one was killed the incident caused major environmental and equipment damage and resulted in the plant being shutdown for more than 6 months. It occurred when furnace tubes became overheated because burners were started using fuel gas with a higher calorific value than normal and with an insufficient flow of process nitrogen through the tubes due to restricted availability. The inexperienced operator did not appreciate the risks of starting the furnace under these conditions.

The investigation highlighted that "the plant, as originally designed, could not be described as inherently safe" [22]. One particular issue that added to the complexity of plant operation and meant operators required a good understanding of its operation and associated risks was that there was no direct measurement of tube skin temperatures provided, even though hazards of tube failure are well known. Indirect indications of temperature had to be used instead. Fundamentally, there was an almost full reliance on operators following start-up procedures to avoid incidents, but the controls required to ensure effectiveness were not in place.

### 3.3.6 Elimination

Elimination involves changing a process design or removing a hazardous activity. The effect is to eliminate the hazardous event completely. Examples include removing a storage tank from a design or avoiding the need for pigging a pipeline [8].

Elimination did not appear in Kletz's list because he generally saw it as a result of applying inherent safety rather than a principle in itself. However, he did make specific reference to elimination when discussing his principles of inherent safety as an option for limitation of effects.

### 3.3.6.1 *Buncefield incident*

On 11 December 2005 an explosion occurred at the Buncefield oil products terminal in the United Kingdom. It was one of the most significant explosions in Europe since the Flixborough disaster.

The explosion occurred because 195 tonnes of petrol (gasoline) leaked from a storage tank when it was overfilled following a transfer from another site. A huge 500 m by 350 m vapour cloud was formed in the zero wind conditions [23]. A tanker driver at the loading gantry reported the cloud, which initiated the site emergency system. The site's fire pump was started, providing a source of ignition. The subsequent vapour cloud explosion caused an overpressure of 2000 mbar [23]. Although there were no deaths or life changing injuries, this incident resulted in £1 billion of physical damage.

One of the causes of the Buncefield incident was that high-level alarms failed to operate. Clearly, an inherently safer design would have reduced the risks without relying on such devices. In this case it would not be practical to stop storing petrol (at least until the internal combustion engine is phased out) but there is scope to eliminate vapour cloud explosions from storage tank overflowing from the roof of storage tanks. The UK Health & Safety laboratory has spent the years since the incident reviewing how to do this [24].

The Buncefield explosion was so large because the overflow from the roof of the tank acted like a 'waterfall', atomising into small droplets which provided more surface area to accelerate the evaporation of the liquid gasoline into vapour. An option being considered to eliminate this is to instal overflow or downpipes near the top of the storage tank. These would mean that liquid from an overfilled tank would be directed to a safe collection area where it can be safely managed and removed. This eliminates the waterfall effect and hence the potential for a significant vapour cloud explosion.

An overflow or downpipe would be both permanent and inseparable from the tank design. This means it would be more reliable than installing additional layers of protection such as additional level switches and valves, which are all prone to failure. Although this still results in a small vapour cloud from the splashing as the gasoline exits the downpipe, the risk that is eliminated is the potential for a significant vapour cloud explosion impacting adjacent populations, and so is inherently much safer.

Downpipes have started to be installed more widely, both on new storage tanks, and also retrofitted to existing storage tanks at different locations in the United Kingdom and around the world. However, it has not been recognised universally across the industry and there is still quite a lot of effort going into installation of safety instrumented devices (e.g. high level instruments which close feed valves) instead.

There was a view shortly after Buncefield that it was a one-off event and so may not warrant too much attention. This was not the case and 5 years later a similar explosion occurred at the Caribbean Petroleum Corporation storage tank facility in Puerto Rico when a tank overflowed during offloading of gasoline from a tanker ship. This caused destruction of 17 storage tanks on the site and damage to buildings over a mile away [25]. Also another similar explosion occurred in Jaipur, India in the same year and further research has shown that incidents had occurred before Buncefield including at Newark, United States in 1983.

### 3.3.7 Segregation

The aim here is to limit the effects of a hazard by providing physical separation between the source of hazards and receptors such as people or the environment [8].

Although segregation did not appear in Kletz's list of inherently safe principles it is clearly related to his second concept of "People who are not there can't be killed."

One of the first options to be considered on any project is where it will be located. This highlights why segregation is a valid principle of inherent safety. For example during a project we can minimise the equipment inventory of a toxic compound, simplify the process, or moderate the reaction conditions to reduce the potential for an exothermic reaction. However, increasing the distance between the source and potential impacted receptors should also be considered and is often the best way to minimise risks.

Segregation can also be applied to newer industries like hydrogen fuelling stations for hydrogen powered vehicles. The UK HSE undertook an experimental demonstration that shows how liquid hydrogen releases from a 1 in. hole can create an explosion and a jet fire. Locating hydrogen fuel pumps on retail sites away from populations, buildings, and other equipment based on the distances in this UK HSE study [26] would be an inherently safer design.

#### 3.3.7.1 *Tianjin explosion*

On 12 August 2015 a series of explosions occurred at the Tianjin factory in the Hebei province of China. One hundred and seventy-three people were killed, including many firefighters. Hundreds of others were injured and hundreds of homes were destroyed [27].

It is believed that that the accident was triggered initially by a fire. Attempts were made to extinguish this fire but 40 min later an explosion occurred, followed shortly after by a much larger explosion. This chain of events occurred because hazardous materials were being stored in close proximity to people and the consequences would have been much less if effective segregation had been applied.

## 3.4 Applying inherent safety through a system's lifecycle

The concept and principles of Inherent Safety can be applied at all stages of a system's lifecycle. However, the greatest opportunities for risk reduction are usually found at the earlier stages of development because there are more options to eliminate or significantly reduce hazards by changing the fundamental design or location. Also, making these changes earlier is likely to be cheaper and cause fewer knock-on issues.

### 3.4.1 Inherent safety in projects

As Kletz said [18],

---

**Text Box 3.27**

If we are to develop inherently safer designs, we need to carry out studies similar to hazard and operability studies (HAZOPs) much earlier in design, at the conceptual stage when we are deciding which process to use, and at the flowsheet stage, before detailed design starts. These studies should be additional to the usual HAZOP of the line diagram and will not replace it.

---

Whilst inherent safety can be applied at all stages in a project, the options change as it develops.

### 3.4.1.1 Inherent safety at the conceptual stage

Kletz acknowledged that what he viewed as the conceptual stage of a project may be defined differently by others. He clarified by saying that this is the stage when decisions are made about what to make, by what route, and where the facility will be located [10]. At the time he started talking about this the idea of considering inherent safety so early in a system's lifecycle was new. It is good to know that since then formal studies at the conceptual stage have become much more common (although still overlooked in many places).

Kletz and Amyotte explained the benefit of such an early start [2].

---

**Text Box 3.28**

The words identify, prevent, control, and mitigate, applied to hazards, are often used to describe the stages to be followed to obtain a safe plant. Inherently safer design could be described as part of prevention, but it is better to add avoid or eliminate to the list of key words, to emphasise its importance. Designers seem to respond to the word control by thinking of added protective equipment; the word does not trigger thoughts of avoidance.

Applying the concept of inherently safer design at the very beginning of a project, we may be able to choose a safe product instead of a hazardous one. When a process is being chosen, we may be able to choose a route that avoids the use of hazardous raw materials or intermediates.

Once the chemistry has been decided and we are developing the flowsheet, we may be able to choose or develop intensified equipment such as reactors, distillation columns, and heat exchangers that do not require large quantities of materials. It may be possible to manage without intermediate storage, possibly by siting production and consuming plants near each other, and to avoid flammable heat-transfer fluids or refrigerants.

---

### 3.4.1.2 Inherent safety at the start of design

Kletz described early design as the 'flowsheet stage'. Today it may be called Front End Engineering Design (FEED) or Define; but one of the outputs is likely to be a flowsheet that identifies the main subsystems and how they are connected.

Kletz and Amyotte provided a comprehensive list of issues to consider before finalising the flowsheet, with the suggestion that these could be included in an early design HAZOP or equivalent study. The issues to consider for a process plant at the early stage of design include [2]:

- Materials—develop an inventory, identify their hazards, and consider options to remove or reduce;

- Reaction—size of reactors and opportunities to reduce; process conditions and opportunities to make less severe; and any potential for runaway reaction;
- Separation—inventory of material in separators and opportunities to reduce;
- Heat transfer—inventory of material in exchangers and opportunities to reduce; use of less hazardous heat transfer medium; ensuring the most hazardous material is in the safest part of the exchanger (e.g. in tubes not shell);
- Storage—factors defining storage requirements and options that would reduce these; storage process conditions and options to make less hazardous;
- Equipment types—options to use simpler alternatives;
- Human error—options to reduce susceptibility that do not involve additional safety systems.

The same questions can be asked throughout the design but it becomes increasingly difficult as the project develops. For example, location options may be very limited very early on, especially if it has taken years to negotiate purchase of new land. Hence, inherent safety should be part of the land selection process.

### 3.4.1.3 Inherent safety during detailed design

Although it may become more difficult it is still important to continue looking for options to increase inherent safety as the design is developed. Examples include eliminating or minimising the stored inventory of hazardous materials, substituting a more corrosion resistant material of construction for equipment, minimising potential hazardous impact by locating access routes and roads away from potentially hazardous areas, locating emergency equipment such as fire water pumps and switch gear for emergency equipment away from the main plant which it is designed to protect, and designing the equipment arrangement in well vented and open process areas to prevent accumulation of hydrocarbon if released.

Even towards the end of detailed design there will still be decisions made that can contribute. Kletz provided a list of details to consider during the HAZOP or other safety studies including [10]:

- Types of gasket;
- Valve design;
- Locations where positive isolation will be required and the methods to be used (e.g. spades, blinds, removable spools);
- Hoses and other flexible connections;
- Sample points;
- Use of quick release couplings;
- Fragile items (e.g. glass, plastic).

As well as selecting items based on safety, the full lifecycle costs and potential complexity should be considered. For example, standardising gaskets, nuts, and bolts can reduce requirements for stock.

### 3.4.1.4 Inherent safety during construction

Although the design is likely to be fixed, there are still opportunities to apply inherent safety when planning for the construction stage of a project. Examples include minimising the hazards of simultaneous operation and construction activities by installing modular units constructed offsite rather than on site construction, and eliminating the potential impact of a blast wave on portable occupied buildings associated with construction by relocating away from potentially impacted areas. Also, elimination of heavy lifts over live equipment, and location of muster points for construction workers in safer areas.

## 3.4.2 Inherent safety during operation

Whilst applying inherent safety at the start of a system's lifecycle provides the greatest opportunities to reduce risk it does not mean that it is only relevant to projects. There can be many opportunities to apply inherent safety during the operational stage of the lifecycle of a system.

Whilst identifying opportunities for inherent safety should be a continual goal, there will be specific times when it should be considered formally. For example, inherent safety should be an option when identifying actions following an incident investigation or when evaluating a plant or process change, instead of simply specifying additional safety systems or changes to procedures. Also, it should be a key part of any routine review of safety studies and of any study carried out retrospectively. In fact, inherent safety was a part of changes to Risk Management Plans (RMP) introduced in 2017 by the United States Environmental Protection Agency (EPA) [28], which apply to petroleum, coal products manufacturing, chemical and paper manufacturing where Process Hazard Analysis (PHA) revalidation is required every 5 years.

Applying inherent safety to an operating system is not straight forward and care has to be taken to avoid unintended consequences. These can include loss of operational flexibility, reduced plant availability, or transfer of safety risks. It is a fine balance to select reasonably practicable options whilst not affecting economic viability.

### 3.4.2.1 Managing inventories

Just because a tank or vessel can hold a quantity of material, it does not always need to be filled to capacity. Reducing inventories to only what is needed will reduce the potential consequences of failure. On the other hand, reduced inventories will inevitably mean that materials need to be handled more often (e.g. smaller deliveries carried out more often). The risk of additional handling needs to be considered against the reduction of risk through reduced inventory.

### 3.4.2.2 Inherent safety during maintenance

It is standard practice to isolate, drain, clean, and purge process equipment before maintenance. But decisions can be made about how much plant needs to be prepared in this way. The inherently safer approach is to shut down and prepare the whole facility because this will minimise the inventory of material present whilst the maintenance is being carried out and also reduces the potential consequence of maintenance errors (e.g. someone breaking the wrong pipework joint). However, it can have significant impact on production. Also, preparing plant and equipment for maintenance, and returning it to service after maintenance carries its own risk and so a balanced approach has to be taken.

Another decision to make when carrying out maintenance is the type of isolation to be used. The inherently safer option is to use positive isolation, with removal of spool pieces being the most robust option because the alternative methods (e.g. spades, blind flanges) are easier to defeat. However, all forms of positive isolation involve breaking joints and so introduce their own risks.

### 3.4.2.3 Inherent safety during hazardous operations

Whilst hazard is present there is always some risk. But certain operations such as plant start-up and shutdown are known to be more hazardous. In these cases the concept of "People who are not there can't be killed" can be applied by clearing sites during the most hazardous operations, or at least limiting them to essential personnel only. Other hazardous operations where this applies include tanker deliveries of hazardous material, opening pig receivers/launchers, sampling, and any activity involving a break of containment, such as filter changing.

### 3.4.2.4 Inherent safety of obsolete equipment

There can be a tendency to allow obsolete equipment to remain on site. This can be because of the expense of removal or 'just in case' it may be used again in the future. This introduces the potential for hazardous materials to collect or hazardous conditions to develop over time. These can either result in incidents in their own right or contribute to the escalation of other incidents.

Ideally, any equipment not being used would be removed. Where this is not possible or desirable it should be fully cleaned and purged, and 'air gaps' created so that there is no connection between the obsolete items and the operating plant.

## 3.4.3 Applying inherent safety in practice

Inherent safety is often considered to be a philosophy: a way of thinking. However, for it to work in practice, inherent safety has to be applied as a systematic process.

Applying inherent safety to complex systems is particularly challenging but it is here that the greatest benefits can be achieved. It is not something that can be done alone, so knowing who to involve is one of the main considerations. In most cases input from different discipline engineers will be required because they all have different expertise and different areas of responsibility.

Collaboration leads to better decision making when applying inherent safety. But like all safety studies a clear process is required to ensure the approach taken is rigorous and systematic. Three different approaches are described as follows [29].

### 3.4.3.1 Focus on high severity hazards

The greatest benefits from applying inherent safety are from eliminating or reducing potential events with high severity (major accidents). The challenge here can be that these events occur very infrequently, making it difficult for people to fully appreciate the benefits when they are having to deal with the frequently occurring lower consequence events.

Potential high severity events will usually be identified in a hazard identification study such as HAZID or a Quantified Risk Assessment (QRA). The requirement to formally apply inherent safety should be identified from this study leading to a separate follow-up review.

The output from the hazard identification study (HAZID or QRA) becomes the input to inherent safety review, but if action is taken that results in an inherently safer solution it may be possible to update the results of the original hazard identification study to show the improvements achieved.

Examples showing where a focus on high severity events can lead to inherently safer solutions include:

- Increasing the design pressure of low pressure equipment to remove any high-low pressure interfaces. Capital costs are increased but the possibility of overpressuring systems is removed;
- Minimising quantities stored of hazardous materials or moving their location away from human populations;
- Moderating temperatures so that the materials being handled are below their flash-point or auto-ignition so that fire does not occur in the event of a loss of primary containment; and
- Removing intermediate storage by using existing capacity in other vessels like the bottom of a distillation tower or reflux drum.

### 3.4.3.2 *Applying inherent safety during option selection*

During the lifetime of a project (including modifications of operating systems) many different options may be studied. Inherent safety should be a key part of selecting the preferred option. One way to do this is to define inherent safety goals and include these in the evaluation process. Conducting a formal and structured Inherent Safety Workshop is recommended, with the focus on changing the basic demand so that add-on safety measures will not be relied on so much [30]. An advantage of doing this is that the project team will improve its understanding of inherent safety principles, which will be of benefit in later stages of the project [30].

A clear definition of inherent safety goals can be used to rank options using a scoring method, which is used to assist decision making [29, 31]. This is best done by a team so the decision-making benefits from the experience and expertise of different discipline engineers.

Inherent safety goals should be selected to be relevant to a project. They may include minimising impact on populations, including third parties offsite, and minimising incidents during construction.

An example of option selection common to most projects is location. The first goal is to select an option away from existing populations. But another consideration is whether populations can be moved, for example by relocating a building or moving the workplace of people currently based there. This can involve a trade-off between everyday convenience and safety, which may not always be obvious to everyone because of the low probability of high severity events.

If a scoring system is used for an evaluation the scores would normally be applied according to how well an option aligns with an inherent safety goal. The range could be between 'fully aligned' and 'no alignment', with a scale to represent partial alignment. The scores may be weighted if the inherent safety goals are not deemed to be equal. Although the selection of the scoring system and the implementation of the process is not scientifically rigorous, the main value of this process is in the way it brings teams together in understanding and identifying hazards and the ways to manage these hazards, so assisting the decision-making process, and in doing so, selecting an inherently safer option.

Examples of option selection include conceptual designs or selection of a technology. The latter may be dependent on the licensor for a process, which can be challenging because the licensor is understandably reluctant to share details of their offering due to commercial confidentiality. Despite this, it still may be possible to get some idea of how well inherent safety goals are being achieved by looking for indications that suggest a reliance on add-on safety systems, for example, the number of relief valves installed, the maximum relief load for flare sizing, or safety instrumented systems with a high safety integrity level. These all indicate that there has been little consideration given to inherent safety in the design.

### 3.4.3.3 Checklists during general engineering design reviews

One issue can be that people do not have the opportunity to formally consider inherent safety on a regular basis. A checklist approach can prompt them to consider the options and prevent items being forgotten [32]. This checklist can be applied as a team to bring in different discipline expertise and experience. However, when using a checklist, it is usually appropriate to remind the team of the inherent safety principles which underpin the checklist, and in this way, build capability and understanding within the team of ISD.

The following is an example of a checklist that can be adopted for this purpose (Table 3.2).

**TABLE 3.2** Example checklist for considering inherent safety at design reviews [29].

| Checklist | Examples of inherent safety for consideration |
| --- | --- |
| Process design | Simplify process design (e.g. fewer processing steps, reduced control complexity) |
| | Reduce equipment count |
| | Increase design tolerances to cater for excursions from normal operation |
| | Minimise line size to limit potential leak rate |
| | Reduce equipment sizes and inventories |
| | Minimise piping lengths carrying hazardous material |
| | Reduce the need for storage and intermediate storage |
| | Avoid sources of ignition by design (e.g. no fired equipment) |
| | Minimise need for transportation of people and hazardous material |
| | Adopt less hazardous transportation methods (e.g. pipeline versus road) |
| Engineering design | Use corrosion/erosion resistant materials |
| | Fewer small-bore connections |
| | Use higher reliability rotating equipment to minimise sparing |
| | Increased component reliability to minimise need for maintenance, disassembly and intervention |
| | Use permanent equipment rather than temporary (e.g. for pigging) |
| | Minimise need for maintenance in hazardous areas |
| | Adopt simplicity in equipment numbering/layout/HMI design to minimise error potential |

*Continued*

**TABLE 3.2**  Example checklist for considering inherent safety at design reviews [29]—cont'd

| Checklist | Examples of inherent safety for consideration |
|---|---|
| | Minimise flanges and unions |
| | Minimise dead legs and low points |
| | Eliminate unnecessary instruments |
| | Bias to nonintrusive instrumentation where practical |
| | Minimise manual sampling/tank dipping/chemical injection |
| | Avoid hot surfaces in hazardous areas |
| | Bias to automation over manual activity |
| | Minimise equipment needing frequent inspection |
| | Provide isolation facilities for safe maintenance, emergency situations |
| | Design for human error tolerance (e.g. unique connections at multiconnection offloading points) |
| | Bias for passive over active protection (e.g. fire protection) |
| | Minimise need for hazardous activity (e.g. pigging, manual handling) |
| Layout/siting | Locate occupied buildings outside hazards zone |
| | Minimise processing and drains systems in enclosed areas |
| | Reduce the size of congested areas (reduce vapour cloud explosion potential) |
| | Separate equipment to minimise potential for escalation |
| | Design to direct liquid spills away from hazardous equipment |
| | Route piping containing hazardous materials away from work areas/vulnerable process equipment |
| | Route vehicle movements away from hazardous areas to reduce pedestrian and vehicle interfaces |
| | Locate critical protective equipment and controls away from process hazards (e.g. firewater pumps/storage) |
| Activity | Avoid exposing construction workforce to hazards of operating plant |
| | Use modular construction to reduce construction workforce exposure |
| | Minimise activities needing confined space entry, working at height, working in inerted spaces, etc. |
| | Minimise need for scaffolding access in process areas |
| | Plan to avoid hot work in process areas |
| | Minimise hot-taps by doing tie-ins during unit outage |
| | Eliminate or minimise SIMOPS (e.g. drilling whilst producing wells) |
| | Minimise need for lifting, elevation, and weights |
| | Minimise temporary clamping |
| | Minimise work on live equipment |

**TABLE 3.2** Example checklist for considering inherent safety at design reviews [29]—cont'd

| Checklist | Examples of inherent safety for consideration |
|---|---|
| Project management | Challenge availability targets where they are causing undue complexity and intervention requirements |
| | Phase project execution to avoid simultaneous construction and operation, to minimise exposure of construction workforce |
| | Set construction and start-up philosophy to avoid construction activities during start-up of adjacent equipment |

### 3.4.3.4 A cultural issue

Applying inherent safety concepts requires hazards to be identified and for people to consider the wider issues. This can be particularly challenging when resources are focused on operational activities and when process safety incidents, particularly major accidents, are so rare as to not be considered.

Integrating inherent safety into decision making is largely a cultural issue. Having a sense of 'chronic unease'—continuously looking for potential issues and aiming to uncover root causes rather than simply fixing the immediate issue—results in people recognising that absence of incidents does not mean that there is no hazard and so inherent safety can always be applied.

Kletz shared the following examples where hard working and diligent people often overlook the opportunities to develop inherently safer solutions because they are too focussed on dealing with the systems or consequences of events instead taking a step back to understand what is happening and how an inherently safer solution could be applied [33].

---

### Text Box 3.29

(a) A cylinder lining on a high-pressure compressor was changed 27 times in 9 years. On 11 occasions it was found to be cracked and on the other 16 occasions it showed signs of wear. No one asked why it had to be changed so often. Everyone just went on changing it. Finally, a bit of the lining got caught between the piston and the cylinder head and split the cylinder.

(b) While a man was unbolting some 3/4 in. bolts, one of them sheared. The sudden jerk caused a back strain and absence from work. During the investigation of the accident, seven bolts that had been similarly sheared on previous occasions were found nearby. It was clear that the bolts sheared frequently. If, instead of simply replacing them and carrying on, the worker had reported the failures, then a more suitable bolt material could have been found.

Why did they not report the failures? If they had reported them, would anything have been done? The accident would not have occurred if the foreman or the engineer, on their plants, had noticed the broken bolts and asked why there were so many.

(c) A line frequently choked, As a result of attempts to clear the chokes, the line was hammered almost flat in several places. It would have been better to have replaced the line with a larger one or with a line that had a greater fall, more gentle bends, or rodding points.

---

Taking the second incident (b) discussed previously as an example, the inherently safer decision would be to replace the bolts with stronger ones. The challenge is that changing all the bolts may have created more work in the short term and so may not have been considered as viable. However, it would have led to a reduction in both work and safety risks over the longer term. One way to integrate inherent safety into day-to-day operations and prevent incidents before they happen is to start asking your front line staff what is making their work difficult. This requires a more searching attitude to finding out about how things work in practice rather than how it may be imagined or described in procedures. But it also requires buy in from front line staff, which requires trust and a demonstrated willingness to respond to make things better.

## 3.5 How far should we go?

Kletz was well aware that inherently safer options can be expensive and cannot always be justified. But he was very keen to emphasise that any decisions made should be based on all the facts and not simplified short-sighted criteria [9].

---

**Text Box 3.30**

First, we have to decide on the level of safety we require. This may he expressed as a Fatal Accident Rate, as the risk to a member of the public living near the factory, as a hazard rate, or in some other way that is considered appropriate.

The two ways of achieving the objective must then be costed—an inherently safer plant (or process) and a conventional or extrinsically safe one with 'added-on' safety. We must remember that we shall probably want to add on more protective equipment than on the last plant, that in some countries the authorities may insist on more protective equipment than we think is necessary, that yet more may be needed during the life of the plant and that all this added-on safety requires testing and maintenance.

We should then adopt the cheaper solution, even if it is the extrinsically safer one. If the two costs are about the same, then we should prefer the inherently safer solution as the 'intangibles' are in its favour. It will be easier to convince the authorities and the public that the plant will not blow up.

The shortcoming of this procedure is that the conventional design may be cheaper because no research has been done on alternatives. It is therefore necessary at corporate level to take a definite decision to carry out research on inherently safer plant. Whether or not we adopt the results of the research should depend on the relative costs of the new and conventional plants, as already discussed.

Once somebody develops an economic inherently safer design, then it will become harder for others to build new extrinsically safer plants for the same product. They can no longer say that they are following 'best current practice' or that the alternatives are not 'reasonably practicable'. Those who are giving no thought to inherent safety had better beware they may be left behind in the race.

---

### 3.5.1 Controlling risks in practice

Although the goal of inherent safety is to avoid reliance on added safety systems, it is recognised that it can never be fully effective. This means safety systems are often required, and in deciding which to implement it is important to recognise that they are not all equally effective or reliable.

#### 3.5.1.1 *Is inherent safety a risk control?*

There can be some confusion because inherent safety aims to avoid a hazard and if avoided it does not need to be controlled. On this basis it is possible to argue that inherent safety is a separate entity. However, pragmatically it is clear that managing risks is achieved in practice in multiple ways and making clear-cut distinctions is neither necessary or helpful.

#### 3.5.1.2 *Hierarchy of risk controls*

The concept of the hierarchy of risk control illustrates that the effectiveness of controls depends on their characteristics. As a general philosophy controls at the top of the hierarchy should always be considered first because they are most reliable. But the overall solution is likely to involve controls at all levels of the hierarchy.

The following is generally considered as an appropriate hierarchy of risk control [8]:

1. Inherent safety—the most effective;
2. Passive controls (e.g. surface coatings, atmospheric vent);
3. Active controls (e.g. relief valve, safety instrumented system);
4. Procedural controls (e.g. start-up procedure, permit to work, response to alarms—least effective).

In addition, mitigation may be considered as an additional control to address residual risks, in order to reduce the consequences of failure of other controls. This includes items such as catchment bunds or dikes, personal protective equipment, and emergency procedures.

It is noted that the US EPA make it clear in their RMP regulations that the term 'hierarchy of control' should not be used and instead state that controls should always be used in the specified order [28]. Interestingly this aligns exactly with the earlier definition of hierarchy of controls, so we have an alignment of the principle even if the EPA does not explicitly use of the term 'hierarchy of controls'.

#### 3.5.1.3 *Defence in depth*

Kletz used the concept of 'defence in depth' to explain the role of inherent safety in the wider safety context. He said [10].

---

**Text Box 3.31**

The prevention of accidents is based on defence in depth. If one line of defence fails, there are others in reserve.

---

These lines of defence are similar in concept to James Reason's Swiss Cheese model [34] in that each line, or layer of protection, prevents events occurring. However, none are fully reliable and failures are represented as holes in the pieces of cheese. When these holes line up a failure pathway is created and results in an accident.

Kletz and Amyotte pointed out that lines of defence act at first to prevent an event, then mitigate to reduce the hazard of an event and finally act to protect people. They listed them as follows for a flammable hazard [2]:

1. Avoidance of hazard (inherent safety);
2. Containment;
3. Detection and warning;
4. Isolation;
5. Dispersion;
6. Removal, as far as possible, of known sources of ignition;
7. Protection of people and equipment against the effects of fire and explosion;
8. Provision of firefighting facilities.

They went on to say [10].

---

**Text Box 3.32**

Similarly, in industry, a common failing is to ignore the outer lines of defence because the inner ones are considered impregnable. When they fail, there is nothing to fall back on. Some companies have not worried about leaks of flammable gas because they had, they believed, removed all sources of ignition so that leaks could not ignite. When an unsuspected source of ignition turned up, an explosion occurred. More often, companies have not worried about large inventories of hazardous materials because they knew, they thought, how to prevent leaks and deal with any that occurred. When they were proved wrong, a fire or explosion occurred or people were killed by toxic gas or vapour.

Effective loss prevention is based on strong outer defences, i.e. on getting rid of our hazards when we can rather than controlling them by added equipment that may fail or may be neglected. It is hubris to imagine we can infallibly prevent a thermodynamically favoured event.

---

## 3.5.2 ALARP

The Acronym ALARP is now widely used and stands for "As Low As Reasonably Practicable." In a safety context it is used when discussing whether a risk is ALARP or further action is required to reduce it.

### 3.5.2.1 Kletz's view of ALARP

Kletz may not have used the term ALARP but he did describe 'reasonably practicable' being relevant because it is impractical to remove every hazard [33]. He implied that the size of a risk and whether it can be tolerated should be compared to the cost, time, and trouble of removing or reducing it.

Kletz recognised that applying an inherently safer solution is not always reasonably practicable and so acknowledged that the principles of ALARP are important when determining how risks are going to be managed.

### 3.5.2.2 ALARP in legislation

The UK Health and Safety regulatory system [35] is 'goal-setting' rather than prescriptive 'rule-setting'. This means it is the duty holder's responsibility to demonstrate that they are managing their risks rather than the regulator's responsibility for saying how risks shall be managed. This approach has ALARP at its heart, relying on a judgement of what is considered 'reasonably practicable'.

Other countries including the United States have avoided adoption of the ALARP principle [28]. One of the reasons is that it is difficult to define what is considered as reasonably practicable for a given circumstance, which makes it difficult to implement when a regulatory approach is rule-based rather than goal-setting. However, recent clarification by the US EPA to the proposed 2017 RMP regulation for petroleum, coal, chemical, and paper manufacturing uses the concept of 'practicability' when applying inherent safety [28]. This recognises that the implementation of the most inherently safer option may not be practicable especially when applying to existing plant. Considerations for this practicability are listed as including timeframe to implement taking into account economic, environmental, legal, social, and technological factors. In this way, there is no absolute requirement to instal the inherently safer solutions, as these should be considered alongside passive, active, or procedural controls as ways to reduce risk.

However, it should be noted that as this book goes to print, the EPA's Proposed Reconsideration Rule rescinded some of the earlier requirements including those associated with ISD [36].

### 3.5.2.3 Applying ALARP in practice

Due to the origin of ALARP in UK legislation, the UK's HSE provides the most guidance on how to apply ALARP in practice. This tends to rely quite heavily on cost–benefit analysis. Whilst this may play a part, it does introduce complexity and relies on quantified data related to risk, which is rarely available in any useful form.

Ironically, HSE has provided some excellent guidance on applying ALARP, but has buried it in some rather obscure references. In particular, deep in their guidance for permissioning within the Control of Major Accident Hazards (COMAH) regulations it states that "ALARP demonstration for individual risks is essentially a simple concept which can be satisfied by the Operator answering the following fundamental questions" [37].

Question 1—What more can I do to reduce the risks?
Question 2—Why have I not done it?

Answers to the first question are qualitative in nature and involve looking systematically at the risks and drawing up, in a proportionate way, a list of measures which could be implemented to reduce those risks.

The answer to the second question may be qualitative or quantitative in nature depending on the predicted level of risk prior to the implementation of those identified further measures. The guidance states that if "it cannot be shown that the cost of the measure is grossly disproportionate to the benefit to be gained, then the Operator is duty bound to implement that measure" [37]. However, there are often reasons to not implement additional measures that are not purely due to financial cost. Risk transferral is very often a factor where a measure to reduce one risk increases another.

In some cases ALARP can mean that an inherently safer solution is not safer overall. For example, choosing to not make a product, to eliminate a hazard, may simply mean that production is moved to another site, possibly in another country. The alternative may apply lower safety standards. Also, risks of transport will have increased. In this case the issue may be moral rather than economic, and there may be an argument to say that such global issues are not necessarily the responsibility of commercial organisations. However, with increased scrutiny from customers of the supply chains of their suppliers it is possible that keeping production local may be the best solution from all perspectives.

ALARP can highlight why inherent safety is preferred to applying add-on safety devices. For example, High Integrity Pressure Protection Systems (HIPPS) are often installed where specification changes occur on a plant so that some plant is rated below the maximum pressure that can be achieved. These are often complex devices that employ multiple instruments acting on multiple valves using a voting system. Testing them can be particularly difficult and time consuming. If this is taken into account the inherently safer option of using fully rated pipework and plant may prove to be cheaper over the full lifetime.

Applying ALARP can challenge views on what is considered to be the inherently safer solution. For example, interlock systems are often specified for pig launchers/receivers that require multiple valve operations to prepare them for opening to insert or remove the pigs. The interlocks are considered necessary to prevent human errors when operating valves. However, they add complexity, which is against inherent safety principles. The reality is that all interlocks can fail or be overridden. The complexity introduced by the interlock makes it difficult for people to understand what they are doing, or more importantly, why. Interlocks should not be considered as an inherently safer solution and require thorough assessment to ensure that they do actually achieve an overall risk reduction.

### 3.5.3 Friendly systems

Inherent safety can appear to be a very technical process reliant on detailed engineering solutions. But Kletz pointed out that thinking about how people are going to interact with a system during its lifecycle can result in subtle design details that can reduce risks during construction, installation, operation, and maintenance. He described this approach as "Making plants friendlier" [10]. He gave some very practical advice and examples, which are summarised as follows.

#### 3.5.3.1 *Avoiding knock-on effects*

"Friendly plants are designed so that incidents do not produce knock-on or domino effects" [10]. This can be achieved by:

- Providing corridors, about 50 m wide, between units;
- Designing and locating equipment so that if their most likely failures occur the consequences are limited. For example, storage tanks are usually built with a weak weld

where the roof meets the walls, so that, if a tank is over pressurised this weld will fail first and, although the tank is damaged, the liquid inside will remain contained; and

- Constructing plants that handle flammable materials outside of buildings.

### 3.5.3.2 *Making incorrect assembly impossible*

"Friendly plants are designed so that incorrect assembly is difficult or impossible" [10]. For example:

- Compressor valves designed so that inlet and outlet cannot be interchanged;
- Relief valves with different sized flanges on their inlet and outlet so that they cannot be installed the wrong way round.

Kletz and Amyotte cautioned that "human beings have demonstrated an ability to defeat even the most inherent of safety features" [2]. Use of unique hose connections is often proposed as a design that eliminates the potential for people to connect to the wrong vessel. A common example is a site that receives deliveries of acid and caustic liquids from road tankers, which can react violently if mixed. However, tanker drivers often have to deal with many different types of connection and so routinely carry adaptors, which rather defeats the effectiveness of this approach.

### 3.5.3.3 *Making status clear*

"With friendlier equipment, it is possible to see at a glance whether it has been assembled or installed incorrectly or whether it is in the open or shut position" [10]. Other examples include:

- Check (nonreturn) valves clearly marked with their direction of flow;
- Gate valves with rising spindles are friendlier because it is easy to determine if they are open or closed;
- Including spectacle blinds in the original design is friendlier because it is easy to see its status whereas if standard blinds or spades are used it may not be immediately clear if one has been inserted or not; and
- Translucent tanks allow level to be seen directly rather than relying on instrumentation (although their use is limited by the materials available).

### 3.5.3.4 *Tolerance of misuse*

"Friendly equipment will tolerate poor installation or operation without failure" [10]. Examples include:

- Spiral wound gaskets leak at a lower rate than fibre if the joint works loose;
- Expansion loops are more tolerant to poor installation than bellows;
- Articulated arms are friendlier than hoses; and
- Pumps that are tolerant to running dry.

### 3.5.3.5 *Ease of control*

"If a process is difficult to control, we should look for ways of changing the process before we invest in complex control equipment" [10]. It should have the following features:

- Be based on use of physical principles rather than relying on added control equipment;
- Be robust so that it will still operate even though there are approximations in the model of the process on which the control system is based;
- Be resilient so that it can be tested and maintained without interfering with operations;
- Respond gradually to process changes rather than rapidly;
- Have wide safe operating limits;
- Have wide margins between control limits and safe operating limits;
- Not rely on very accurate control; and
- Avoid positive feedback within the process, for example avoiding exothermic reactions that increase their rate when they get hot as this can result in a runaway.

It has been noted that computer control introduces a number of additional issues and work is required to make it as friendly as possible. This includes making it easier to scrutinise and subjecting it to safety studies (e.g. CHAZOP).

### 3.5.3.6 *Instructions and other procedures*

"We should aim for safety by design rather than procedures when it is practicable and economic to do so" [2]. Procedures will always be required but they are often too long and complex. One reason for this is a desire to cover every conceivable eventuality; sometimes due to the writer wishing to protect themselves from criticism instead of focusing on the users' needs. It is better to write an instruction that covers most of the circumstances that can arise and is short and simple enough to read, understand and follow [2].

### 3.5.3.7 *Lifecycle friendliness*

"We should, when designing a plant, consider the problems of those who have to construct and demolish it, as well as those who have to operate it" [2]. Kletz gave examples of having to lift a reactor over existing plant that had to be shutdown, which could have been avoided by changing the design. Also, the issues experienced in the nuclear industry where decommissioning is proving to be difficult because it was not considered during the original design.

The following summarises the role of inherent safety in minimising risks over the full lifecycle of a system [38].

> Inherently safer technologies implemented in the design phase do not create environmental problems, and inherently cleaner technologies do not create safety problems. Instead of resorting to end-of-pipe solutions, inherently safer and cleaner approaches during design and other phases of a plant's life cycle can eliminate potential conflict between environmental initiatives and safety activities.

## 3.5.4 Global issues

The accident at Bhopal has provided us with one of the most graphic examples of why inherent safety is so important. Unfortunately, project design teams are still missing opportunities to design out hazards and continue to rely only on add-on safety devices to reduce risks [30].

The fundamental issues that resulted in the Bhopal plant being built in India remain and there is a continued trend to moving bulk chemicals production and handling to developing countries [39]. Even when the same safety standards are applied to these plants as in other parts of the world the hazards are often greater because plants are larger, and legislation is often less effective at ensuring safety standards are maintained for the lifecycle.

A design exported from developed to developing countries is often justified 'because it works', with minimal consideration of safety issues. "Perversely, it seems that widespread adoption of inherently safer design, the best available technique for reducing risk to people and the environment, has been prevented by inherent aversion to commercial risk and conservatism of the chemical industry" [39]. This can be exacerbated by project evaluations being focussed on economic and logistical considerations at the expense of any serious engineering or safety evaluation.

The message for industry is that it should "Export inherent safety not risk" [39].

# References

A Note about Kletz's books on inherent safety.

Three of the references below have different titles but are essentially different editions of the same book, as follows:

- Cheaper, Safer Plants or Wealth and Safety at Work published by The Institution of Chemical Engineers in 1984 [9];
- Plant Design for Safety—a User-friendly Approach published by Hemisphere Publishing Corporation in 1991 [10];
- Process Plants—A Handbook for Inherently Safer Design, published by Taylor and Francis with the first edition in 1998 and second edition co-authored by P Amyotte published in 2010.

[1] T. Kletz, What you don't have, can't leak, Chem. Ind. 9 (1978) 287.
[2] T. Kletz, P. Amyotte, Process Plant, A Handbook for Inherently Safer Design, second ed., Taylor & Francis, 2010.
[3] T. Kletz, By accident … a Life Preventing them in Industry, PFV Publications, 2000.
[4] Center for Chemical Process Safety, Guidelines for Risk Based Process Safety, Wiley & Sons, 2007.
[5] Center for Chemical Process Safety, Inherently Safer Chemical Processes: A Life Cycle Approach, second ed., (2009).
[6] Center for Chemical Process Safety, Final Report: Definition of Inherently Safer Technology in Production, Transportation, Storage, and Use, American Institute of Chemical Engineers, 2010.
[7] T. Kletz, L. Poulter, D. Mansfield, Improving Inherent Safety, Health & Safety Executive, 1996. report OTH 96 521.
[8] Energy Institute, Guidance on Applying Inherent Safety in Design: Reducing Process Safety Hazards Whilst Optimising Capex and Opex, second ed., (2014).
[9] T. Kletz, Cheaper, Safer Plants or Wealth and Safety at Work (Notes on Inherently Safety and Simpler Plants), The Institution of Chemical Engineers, 1984.
[10] T. Kletz, Plant Design for Safety—A User Friendly Approach, Hemisphere Publishing Corporation, 1991.
[11] T. Kletz, Lessons from Disaster, Institute of Chemical Engineers, 2003.
[12] T. Kletz, Inherently safer design—its scope and future, Process Saf. Environ. Prot. vol. 81, (Part B) (2003) 401–405.
[13] T. Fishwick, Hazards of ammonia, Loss Prev. Bull. 242 (2015).
[14] S. Gakhar, S. Rowe, Runaway chemical reaction at corden pharmachem, cork, Loss Prev. Bull. 237 (2014).

[15] C.A.R. Hoare, The 1980 ACM Turing Aware Lecture, Communications of the Association for Computing Machinery (ACM), 1981.

[16] Chemical Safety Board, Investigation Report—Static Spark Ignites Flammable Liquid during Portable Tank Filling Operation, CSB, 2008. No. 2008-02-I-IA.

[17] Chemical Safety Board, Chevron Richmond Refinery Pipe Rupture And Fire, CSB, 2015. Report No. 2012-03-I-CA.

[18] T. Kletz, Learning from Accidents, third ed., Gulf Professional Publishing, 2001.

[19] Chemical Safety Board, Investigation Report—Catastrophic Rupture of Heat Exchanger (Seven Fatalities) Tesoro Anacortes Refinery, Report 2010-08-I-WA (2014).

[20] A. Reza, E. Christiansen, A case study of an ethylene oxide explosion in a sterilisation facility, Loss Prev. Bull. 235 (2014).

[21] Chemical Safety Board, Investigation Report—Sterigenics, CSB, 2006. Report No.2004-11-I-CA.

[22] R. Prior, PSM lessons from a major reformer furnace failure, Loss Prev. Bull. 237 (2014).

[23] G. Atkinson, J. Hall, A. McGillivray, Review of Vapour Cloud Explosion Incidents, Health & Safety Executive Research, 2017. Report RR1113.

[24] G. Atkinson, Design of Overflow Piping to Mitigate the Consequence of Gasoline Overfilling Incidents, Health & Safety Laboratory, 2013. Report number FP/12/37, Health & Safety Executive.

[25] U.S. Chemical Safety and Hazard Investigation Board, Final Investigation Report: Caribbean Petroleum Tank Terminal Explosion and Multiple Fires, CSB, 2015 Report No. 2010.02.I.PR.

[26] J. Hall, Ignited Releases of Liquid Hydrogen, Health & Safety Executive, 2014. Report number RR987.

[27] Wikipedia, Tianjin Explosions, Accessed from: https://en.wikipedia.org/wiki/2015_Tianjin_explosions, 2015 November 2019.

[28] EPA regulation 40 CFR Part 68, Accidental release prevention requirements: risk management programs under the clean air act, Fed. Regist. 82 (9) (2017). Friday, January 13, 2017/Rules and Regulations, https://www.govinfo.gov/content/pkg/FR-2017-01-13/pdf/2016-31426.pdf. (Accessed February 2020).

[29] C. Skinner, Inherently Safer Design: it's not just what you do it's the way that you do it, and that's what gets results, Loss Prev. Bull. 262 (2018).

[30] G. Ellis, Are we doing enough to reduce hazards at source? Loss Prev. Bull. 240 (2014).

[31] D. Edwards, J. Foster, D. Linwood, M. McBride-Wright, P. Russell, Inherent Safety: It's Common Sense, Now for Common Practice!, Hazards 25, IChemE, 2015.

[32] A. Gawande, The Checklist Manifesto: How To Get Things Right, Metropolitan Books of Henry Holt and Company LLC, 2009.

[33] T. Kletz, What Went Wrong? fifth ed., Elsevier, 2009.

[34] J. Reason, Organizational Accidents Revisited, Taylor Francis Group, 2016.

[35] Health & Safety Executive website, http://www.hse.gov.uk/risk/theory/alarpglance.htm. (Accessed February 2020).

[36] United States Environmental Protection Agency, Prepublication Copy Notice: Accidental Release Prevention Requirements: Risk Management Programs under the Clean Air Act, Docket No.: EPA-HQ-OEM-2015-0725, https://www.epa.gov/sites/production/files/2019-11/documents/prepublication_copy_rmp_reconsideration_finalrule_frdocument_signed2019-11-20.pdf, November 20, 2019. (Accessed February 2020).

[37] Health and Safety Executive HID CI5A, Guidance on ALARP Decisions in COMAH, SPC/Permissioning/37. Version 3, http://www.hse.gov.uk/foi/internalops/hid_circs/permissioning/spc_perm_37/. (Accessed November 2019).

[38] M. Mannan, J. Baldwin, Inherently safer is inherently cleaner: A comprehensive design approach, Chemical Process Safety Report, (2000).

[39] D. Edwards, Export inherent safety—not risk, Loss Prev. Bull. 240 (2014).

# Managing maintenance risk

## 4.1 Introduction

Kletz provides a clear explanation in his autobiography [1] about why maintenance risks were a topic of particular concern to him. He refers to a fire experienced by ICI in 1967 in which a man died. This led him to review the systems in place to manage maintenance risks and he found many issues. This accident and Kletz's findings are discussed in detail in this chapter.

Risks from maintenance activities continue to be an area of concern. A report from the US Chemical Safety and Hazard Investigation Board published in 2017 [2] stated that of the incidents which the organisation has investigated, 37% occurred in chemical and manufacturing facilities prior to, during, or immediately following maintenance activities and resulted in 86 fatalities and 410 injuries. The same report highlighted that incidents are "due mostly to preparing the equipment for the maintenance, rather than the maintenance itself" [2].

Maintenance is not (or at least should not be) a normal mode of operation for any system. It often involves tools and processes that introduce new hazards to site. Also, it tends to disturb the controls that we normally have in place to control hazards.

Particular concerns for the process industry are the potential for air to mix with flammable substances and for substances harmful to people and/or the environment to be released [3]. In some cases hazards come from sources that are not normally a problem. For example, pyrophoric materials such as iron sulphide, which can form when steel is exposed to sulphur, are not normally of concern during operation but create a significant fire risk when exposed to air, which can occur during maintenance [3].

Many of the issues identified with maintenance risks are related to human error and human factors, which are covered in more detail in Chapter 6. However, it is clear from his books that Kletz recognised maintenance as a particular concern that requires far more attention than it often receives. It is a time of change both in terms of the status of the system and the people involved. Maintenance personnel (including contractors) are not always as familiar with the process risks and controls as the system's operators who usually have full control. Maintenance often requires the system to be shutdown before, and started-up afterwards, which are the most risky things we ever do on process plants.

115

### 4.1.1 A necessary evil?

In an ideal world nothing would break or wear out and we would not have to maintain anything. In the real world maintenance is necessary. In some cases we wait for something to break and then fix it. In others we carry out preventative maintenance to reduce the likelihood of breakdowns.

Maintenance does not always receive the attention it deserves. During design, the trade-off between 'Capex and Opex' (capital expense vs operational expense) leads to cheaper options being chosen that keep the initial cost low even though everyone knows the total lifecycle cost is going to be higher due to higher maintenance costs. At one company it was well known that projects always specified a certain type of relief valve because they were cheap but within a year or two they would always be replaced with a more expensive alternative because operations demanded higher reliability.

Whilst decisions based on cost can be a little perverse, Kletz pointed out a more significant problem [4].

---

**Text Box 4.1**

Maintenance is a hazardous activity. Many accidents occur during maintenance, often as a result of poor preparation rather than the maintenance itself. One way of reducing maintenance is to avoid, when possible, equipment with moving parts.

The less maintenance we carry out, the fewer accidents we will have. As a first step in reducing maintenance, we need to know how we are spending money at present. Many maintenance cost accounting systems are designed to allocate the costs to the right product and cannot, for example, provide data on the costs of maintaining different sorts of pumps or even the costs of different sorts of work.

---

Requiring design teams to make choices based on full life costs and risks would inevitably result in increased upfront costs and effort. But is it really realistic to expect this to happen? Kletz certainly thought so and used the nuclear industry as an example [4].

---

**Text Box 4.2**

The nuclear industry does not achieve high reliability by massive duplication, by making everything thicker and stronger, or by using expensive materials of construction or special types of equipment; rather, it achieves high reliability by paying great attention to detail in design and construction. The last item is important. Many failures in the process industries are the result of construction teams not following the design in detail or not carrying out well, in accordance with good engineering practice, details that were left to their discretion.

---

### 4.1.2 What counts as maintenance?

We carry out maintenance to keep our systems working as they should. When something breaks it is fairly obvious it needs to be repaired or replaced, and this activity is described as maintenance.

We perform a range of other activities that also contribute to keeping things working. This often involves inspecting and testing items to confirm they are still working as required, either to identify deterioration that increases the likelihood of failure in service or unrevealed faults that are not obvious from the available operational data. Cleaning is another action that can be included within the maintenance scope.

We are increasingly relying on software within electronic systems, both in an operational and safety capacity. Although software does not degrade in the same way as hardware (i.e. it doesn't wear out or spontaneously break) it often requires maintenance to correct faults, respond to changing demands and threats from viruses and other cyberattacks, and to implement improvements. All of these result in changes to the software and if there are any safety implications they need to be subject to robust management of change.

### 4.1.3 Maintenance modes of operation

Most maintenance creates some disturbance to the operation of a system and so represents a different mode of operation. The only exception may be external visual inspections where the system can be in any state including its normal operating mode.

Part of managing the risks of maintenance is making sure the system is in an appropriate mode of operation. The options include:

- System is in operation;
- Operation has been paused but process materials remain in place;
- Operation has been stopped and the system being maintained has been made safe by removing all hazardous materials and conditions. However, adjacent systems remain in operation;
- The whole site has been shutdown. This is often described as a 'turnaround' or 'outage'.

It may appear that the safest option is always to shutdown and make the system safe before any maintenance can take place. However, shutting down and subsequent restart carries its own risk and a disproportionate number of major industrial incidents occurs during these activities [5]. Also, some maintenance (particularly testing) requires systems to be operational to be effective.

The perceived status of the system can have a significant impact on how people carry out maintenance. People are far less likely to take precautions if they believe it to be safe but if they work on a live system they are likely to be far more cautious.

As an example, live-line electrical work is quite common in some places, including on high voltage systems. At first it can appear to be a very unsafe thing to do but it actually has a good safety record. To understand this apparent contradiction it is important to note that lots of people have died whilst working on electrical systems that they believed to be 'dead' (isolated and de-energised). This happens because people isolate the wrong system or isolate the right system but then work on the wrong one. The other cause is the system is accidentally re-energised whilst the work is taking place.

If you know you are working on a live system the hazard is very obvious, the safety controls are very clear and well known, and everyone understands exactly why the controls need to be followed. In theory the safest approach would be to isolate the system but for work to be carried out as if it is live. However, human behaviour means that people quickly relax their guard because they know the system is safe meaning this approach is rarely effective.

### 4.1.4 Typical maintenance process

It is relatively easy to map a typical maintenance process. First the system has to be shut-down, it is then isolated from sources of hazard and energy, residual hazards and energy are removed, and then maintenance can take place. When completed the system can be restarted and returned to service.

Unfortunately the actual process can be far more complicated. It is often not possible to separate the different stages completely because the way one is performed will affect another. For example, the method used to shutdown the system can affect the ability to isolate it or remove residual hazards.

Examples of why maintenance is often more complex than expected include:

- If a system is shutdown for maintenance the opportunity is usually taken to complete more than one piece of work meaning more people are involved and effort is required for coordination;
- Multiple parties can be involved including contractors, vendors, delivery drivers, etc.
- Part or all of the system may be live whilst maintenance is carried out meaning operations and maintenance personnel have to be careful that their activities do not interfere with the others.

### 4.1.5 Problems with maintenance

Kletz identified one particular accident that occurred during his time at ICI that alerted him the potential risks of maintenance [1].

---

**Text Box 4.3**

About a year before I was appointed three men had been killed in a serious fire on North Tees Works. A fitter was Opening up a pump when hot oil, above its auto-ignition temperature, came out and caught fire. The plant was destroyed. Examination of the wreckage showed that the suction valve on the pump had been left open. The supervisor who issued the permit-to-work said that he had inspected the pump and found the valve already shut, the pump having been left standing ready for repair over the weekend. Either the supervisor's recollection was incorrect, and he had not checked the valve, or after he had checked it, someone else opened it. Either way, the system of working was poor and after the fire instructions were issued that in future:

- All valves isolating equipment under repair must be locked shut with a padlock and chain (or equally effective means).
- In addition, equipment under repair must also be isolated by slip-plates (spades) or physical disconnection unless the job to be done was so quick that fitting slip-plates would take as long and be as dangerous as the main job. Valves must be locked shut while slip-plates are fitted.

In North Tees Works the fire had been so traumatic that everyone accepted these new requirements without hesitation, but were they really being followed in the rest of the Division? Was it possible to follow them? Were they followed elsewhere? And would they be followed in a few years' time when memories had faded?

---

## 4.1.6 Typical problems

We have a lot of evidence to show that maintenance carries risks in most industries. Understanding why can help us implement more effective controls. A study of major accidents in the metal industry identified the following common problems with maintenance [6]:

- Deficient planning/scheduling/fault diagnosis;
- Deficient preparation for maintenance;
- Deficient performance of maintenance work (based on the maintenance work process) and lack of barrier maintenance and maintenance being an initiating event for an accident scenario, deficient risk assessment;
- Deficient regulatory oversight, deficient implementation of requirements;
- Deficient design/organisation/resource management;
- Deficient documentation;
- Deficient learning (based on the accident process).

## 4.2 Identifying maintenance requirements

All hardware can fail. We have choices to make about whether to wait for this to happen and act reactively to fix it or whether to take proactive action to try and prevent it. Failure to carry out maintenance when it is required or carrying out the wrong maintenance can mean that our risk controls are degraded and can result in hazardous failures.

Deciding when to carry out maintenance or determining what maintenance is required can require a high degree of skill and is prone to human error. This is not always recognised and people can often apply simple approaches to address complex problems.

### 4.2.1 Breakdown maintenance

If something breaks it seems reasonable to assume that it will be repaired or replaced quickly and efficiently. However, this is not always the case for a number of reasons.

#### 4.2.1.1 Faults that develop gradually

Faults that develop gradually or result in a loss of performance rather than a total failure are difficult to detect. The indications are subtle and people quickly become accustomed to the signs of the gradual degradation and so do not recognise it as a problem.

We are all probably accustomed to this with our cars. We may drive it one day and notice a very quiet rattle and make a mental note to pop into the garage when we get the chance to have it looked at. But after a couple of journeys that rattle has become part of the background noise that we consider to be normal and we forget to get it checked. The next time we think about it is when a fault is picked up at the next routine service or the car breaks down on a quiet road, late in the evening in a place with no phone signal.

Fans on computers are good example of faults that develop gradually. As a visitor to a process control room it is quite common to detect a fairly loud and sometimes unpleasant background whine. When asked, the operators who spend hours in there are completely unaware. It can take some time to identify the cabinet where the noise is coming from. This creates two issues. The first is that the background noise can be fatiguing for the operator and potentially interfere with communications. The other is that computer systems rely on fans for cooling and when this is lost it can cause failure of critical operations or safety system, and in extreme cases the loss of cooling can result in a fire.

In October 2003 an employee was killed and two were burned due to an aluminium dust explosion at the Hayes Lemmerz International-Huntingdon Inc. facility, Huntingdon (USA) [6]. Whilst starting up a scrap re-melting system a crust formed in the vortex of a processing machine, which led to chips overflowing into the dust duct, where a fire started leading to the explosion. Excess dust release had been known about for some time but it had not been properly investigated and the system was kept in operation. It is tragic the number of times people are harmed in accidents where earlier warning signs had been overlooked.

### 4.2.1.2 *Faults on spare equipment*

Reliability and criticality studies carried out during design can often result in some items of equipment being duplicated so that one can operate as 'duty' and the other is available as 'spare'. One problem with this arrangement is that repairing a faulty item is perceived as a lower priority because production continues using the back-up spare. People fail to fully appreciate the impact of a subsequent second failure on production and possibly safety. This highlights why philosophies adopted during design need to be understood by people involved in operations. In 1988 a fire and explosion on the Piper Alpha oil rig happened after operations tried ro reinstate the duty pump (isolated for maintenance) after the standby failed. See Section 8 of Appendix.

Another issue with maintaining spare equipment is that it usually involves working alongside operating equipment. It creates a high reliance on robust isolations and requires people to work in hazardous areas.

### 4.2.1.3 *Operating with degraded safety systems*

A lot of effort is put into designing safety systems that achieve a required or target reliability (i.e. they will activate reliably on demand). This is particularly the case with the ever increasing introduction of Safety Instrumented Systems (SIS).

Design engineers tend to have very high regard for engineered safety systems whilst viewing operational controls including response to alarms and operator vigilance as very unreliable. But SIS and other hardware systems can degrade, typically due to component failure. Operators then have to decide whether to shutdown until the system has been repaired or use an override, inhibit, or bypass to allow operations to continue. It is too simplistic to say a system must always be shutdown if its safety systems are not functioning. Equally, it is clearly very risky to continue operating for extended periods of time without functioning safety systems. Companies need to recognise that this situation will arise and provide appropriate support to allow people to make objective assessments of the overall risks when they decide how to respond. They should take into account the impact of the system failure, the effectiveness of other controls, including operational controls, and the options available to improve effectiveness for the duration until the system can be returned to service.

## 4.2.2 Preventative maintenance

We carry out preventative maintenance on apparently healthy equipment to reduce the likelihood of it failing when we need it. Preventative maintenance is essential for items that are critical to production or safety. We usually set the frequency according to calendar time (e.g. annual maintenance), running hours, number of starts, or based on the results of condition monitoring.

### 4.2.2.1 *What to maintain*

Kletz clearly recognised the importance of preventative maintenance of protective equipment [7].

---

### Text Box 4.4

All protective equipment, and all equipment on which the safety of the plant depends, should be tested or inspected regularly to check that it is in working order and is fit for use. If we do not do so, the equipment may not work when required. The frequency of testing or inspection depends on the failure rate. Relief valves are very reliable; they fail about once per 100 years on average, and testing every 1 or 2 years is usually adequate. Protective systems based on instruments, such as trips and alarms, fail more often—about once every couple of years on average—so more frequent testing is necessary, about once per month. Pressure systems (vessels and pipework) on non-corrosive duties can go for many years between inspections but on some duties they may have to be inspected annually, or even more often.

---

Preventative maintenance of the more sophisticated engineered controls such as relief valves and SIS is widely recognised and followed. However, Kletz pointed out that lots of items on process plant can provide a protective function and should also be subject to routine maintenance, inspection, and testing [7].

---

### Text Box 4.5

The following sorts of protective equipment should also be tested or inspected regularly, though they are often overlooked even by companies which conscientiously test their relief valves. In some cases, times of response should be checked:

- Drain holes in relief valve tailpipes. If they choke, rainwater will accumulate in the tailpipe.
- Drain valves in tank bunds. If they are left open the bund is useless.
- Emergency equipment such as diesel-driven fire water pumps and generators.
- Earth connections, especially the moveable ones used for earthing road tankers
- Fire and smoke detectors and fire-fighting equipment.
- Flame arrestors.
- Hired equipment. Who tests it—the owner or the hirer?
- Labels are a sort of protective equipment. They vanish with remarkable speed, and it is worth checking regularly to make sure that they are still there.
- Mechanical protective equipment such as overspeed trips.

- Nitrogen blanketing (on tanks, stacks and centrifuges).
- Non-return valves and other back-flow prevention devices, if their failure can affect the safety of the plant.
- Open vents. These are in effect relief devices of the simplest possible kind, and should he treated with the same respect.
- Passive protective equipment such as insulation. 1f 10% of the fire insulation on a vessel is missing, the rest is useless.
- Spare pumps, especially those fitted with auto-starts.
- Steam traps.
- Trace heating (steam or electrical).
- Valves, remotely operated and hand-operated, which have to be used in an emergency.
- Ventilation equipment.
- Water sprays and steam curtains.

All protective equipment should be designed so that it can he tested or inspected, and access should be provided. Audits should include a check that the tests are carried out and the results acted on.

One item that is often overlooked is maintenance of manual valves, especially ones that are used relatively infrequently. Operators often have concerns that they may hurt themselves when operating stiff valves. But Kletz points out the risks can be more significant than purely to the individual [8].

## Text Box 4.6

Incidents have occurred because valves which have to be operated in an emergency were found to be too stiff. Such valves should be kept lubricated and exercised from time to time.

Clearly preventative maintenance of all the items listed previously can create a significant workload and unfortunately it is one of the things that is often allowed to slip due to lack of staff or other work that is perceived to be more important. Part of the problem is that putting off preventative maintenance rarely has an immediate effect but it can quickly become normal to skip these tasks. The result is that multiple items start to degrade meaning the overall basis of safety is degraded. It can be driven by the attitude summed up in the common idiom "if it ain't broke don't fix it".

### 4.2.2.2 Determine schedules

Maintenance schedules should be determined according to failure mechanisms and likelihood, and the maintenance tasks that can be performed to reduce the associated risks. Methods such as Reliability Centred Maintenance (RCM) and Reliability Based Inspection (RBI) can assist.

If an item can fail during operation due to wear and tear the aim is to schedule preventative maintenance so that the cost of carrying out the maintenance, including loss of production, is less than the cost of failure. However, this requires continuous review to account for changes in the way systems are operated and learnings about the actual reliability of equipment.

In 1999 four workers were killed at Tosco's Avon Refinery in Martinez, California (USA) [9]. The accident occurred when pipework was being replaced whilst the plant remained in operation. The pipework and valves had been 'run to the point of breakdown due to corrosion, leading to a potentially hazardous situation'. The cause of the excessive corrosion was operating an upstream crude desalter beyond its design limits. This resulted in corrosive material being carried over into downstream pipework. However, the change of operation had not been assessed and hence inspection and maintenance schedules had not been modified.

If the concern is that an item can fail to operate on demand the aim is to carry out maintenance, including testing, to confirm its availability is acceptable. The focus should be on 'unrevealed failures', which are not clearly indicated whilst the system is in its 'normal' mode but may mean it will fail to operate on demand leading to a hazardous condition or failed mitigation.

### 4.2.2.3 *Monitoring backlog*

Although there are methods to determine preventative maintenance strategies, to be effective it is necessary to have the right culture in place based on an understanding of how schedules have been developed. Without this there can be a tendency for people to do what they think is appropriate, often based on past experience. This can result in maintenance being carried out more frequently than required, which may actually increase overall risks. Alternatively, it can mean that preventative maintenance is delayed because people do not understand the hidden risk.

Adherence to preventative maintenance schedules should always be a key performance indicator (or similar) and the data should be used to review staffing levels, work organisation, and equipment reliability.

### 4.2.2.4 *Ensuring the correct methods are used*

It is usually easy to confirm that the correct reactive maintenance has been carried out because an observed fault has been rectified. This is not the case for preventative maintenance and it is quite common for people to assume that they are doing what is required when in fact their actions are not addressing the main concerns. This is particularly the case for testing and inspection where people may be reassured that they are not finding any faults but the reality is that they are looking in the wrong place, the method that they are using is not suitable, or they misinterpret the results that they receive.

Kletz gave us an example of how failure to define and/or follow a good procedure can mean testing of a SIS is ineffective [7].

---

Text Box 4.7

The test is not thorough

A trip has three parts: the measuring device or sensor, the processing unit (which can be a microprocessor or conventional hard-wired equipment) and the valve. All three should be tested. The sensor should be tested by changing the pressure, level or concentration to which it is exposed. (If temperature is being measured a current may be injected from a potentiometer.) The valve should, if possible, be tested by closing (or opening) it fully. If this upsets production it may be closed halfway and tested fully during a shutdown. Sometimes, however, the valve is not tested; sometimes only the processing unit is tested; sometimes the sensor is tested but the level at which it operates or its speed of response is not checked.

Kletz highlighted that testing to confirm that a system functions as designed is not just concerned with identifying component failures. Issues can be the result of unplanned or unauthorised changes [7].

**Text Box 4.8**

The setpoint has been altered

Set points should be changed only after authorisation in writing at an agreed level of management.

The trip has been made inoperative (disarmed)

Sometimes this is done because it operates when it should not and upsets production; sometimes it is done so that it can be tested or repaired or because a low flow or low temperature trip has to be isolated during start-up. A turbine on a North Sea oil platform was routinely cleaned with high pressure water. Chemicals in the water affected the flammable gas alarms, so they were isolated. When cleaning was complete the alarms were not reset. When a flameout occurred it was not detected, gas continued to build up in the combustion chamber and an explosion occurred, followed by a fire. Managers and supervisors should carry out regular checks. If a by-passed trip causes an accident, it has probably been by-passed for some time or on many occasions, and an alert manager or supervisor could have spotted it. If a trip has to be isolated for start-up, the isolator can be fitted with a timer so that the trip is automatically re-armed after an agreed period of time.

### 4.2.3 Diagnosing faults

Studies have shown that people are generally quite poor at fault diagnosis [10]. This can mean that maintenance takes a lot longer because lots of different solutions are tried before finding the right one and it can mean that systems are returned to service still in fault, which can contribute to accidents.

Three workers were killed in 2010 at AL Solutions Inc., New Cumberland, WV (USA) [6]. The investigation concluded that sparks or heat had been created by metal blades on a blender contacting metal sidewalls. This ignited zirconium dust and caused an explosion. The blender had undergone maintenance earlier but had been incorrectly returned to service.

One of the reasons diagnosing can be difficult is that the information available about the fault is subtle or apparently contradictory. This means people approach it without a clear idea of what to do and adopt a 'trial and error' approach. This leads them to assuming the problem is the same as previously and they repair or replace the components that usually fail. Unfortunately, because they want that to be the solution they unconsciously look for evidence to confirm that their action has been successful. This is particularly problematic if there happen to be two unrelated faults on the system at the same time.

Modern systems are creating another potential problem with fault diagnosis because they can appear to give very strong indications of the cause of a problem in the form of a fault code or error message. There are no issues if these are accurate but that is not always the case, again especially if there is more than one unrelated fault. The information provided by a system should only be used as one piece of evidence to explain the cause of a problem and it is clear that all other information should also be considered. But when you have a digital read-out giving an apparently unambiguous message it is very difficult for anyone to consider "could it be wrong?"

Improving diagnosis is inevitably reliant on training of personnel to be systematic in their approach, to collect and analyse all of the information available, and to keep an open mind and to consider every eventuality. Procedures can have some use if they support an analytical approach but simple fault-cause-solution prompts have their limitations because they cannot cover every eventuality or address unique problems.

## 4.3 Preparing for maintenance

The safest approach may appear to be to only carry out maintenance on systems that are shutdown and after all hazards have been removed. However, failure to prepare systems properly is one of the main sources of risk and so the 'inherently safe' option may not actually be safest overall.

### 4.3.1 Plant status for maintenance

One of the main decisions to make when preparing for maintenance is to decide if the system needs to be shutdown or whether the work can be carried out with it in its operational state.

#### 4.3.1.1 Shutting down for maintenance

Systems usually need to be shutdown before carrying out any intrusive maintenance, and it is preferable to do so for non-intrusive maintenance as well. It is reported that "more than 20% of incidents occur during shutdown or start-up" [3], which is significantly disproportionate given the relatively short amount of time most systems spend in this mode of operation.

'Normal' operational shutdown procedures can sometimes be followed but in many cases preparing for maintenance requires a different approach, with additional steps.

For example, minimising inventories as part of the shutdown will reduce the time and effort required for draining. Also, it can reduce the quantity of material sent to waste.

Shutting down for maintenance can also provide an opportunity to do things that are not always possible such as testing emergency equipment. Doing this allows faults to be identified so that repairs can be included in the maintenance scope whilst the system is shutdown [3].

It is usually the operations teams that prepare systems for maintenance. They do this frequently and so view it as a routine activity. However, specific requirements can vary considerably and so preparation should be viewed as a non-routine mode of operation and should receive more attention [2].

One reason for developing specific shutdown procedures when preparing for maintenance is the requirement to prove isolations. Things to be aware of include making sure pressure is available at the right time to prove the integrity of every valve being relied on as an isolation and making sure valves are not operated after their integrity has been confirmed.

Major turnaround or outage shutdowns will usually include a different scope each time they are carried out. People tend to copy the previous shutdown procedure and amend it to suit. This can quickly become complicated because people have to decide which sections can stay and which have to be changed. It can be useful to have a single, standard shutdown procedure that is referred to each time and updated to reflect the current scope. Each bespoke procedure is then used once and archived and the 'standard' one is used again next time.

Good practice is to ensure that a written procedure in the form of a checklist is always printed and followed every time a shutdown is carried out, with spaces provided for times and initials to be recorded at every step and sign-off boxes at key stages [3].

### 4.3.1.2 *Carrying out maintenance on live systems*

Whilst inherent safety should always be considered in everything we do, shutting down a system has its own risks and so may not be the safest option. Also, some maintenance can require the system to be in an operational state. Whenever maintenance is carried out on a live system particular attention must be given to ensure risk controls are implemented rigorously and effectively.

In 1989 a fire and explosion at the Phillips Petroleum (now known as Phillips) chemical plant in Pasadena, Texas (USA) killed 23 workers [11] (see Section 11 of Appendix). It happened on a plant producing high-density polyethylene (HDPE). Raw materials were introduced into loop reactors to form polyethylene, which collected in settling legs as 'fluff'. These were prone to blocking so routine, intrusive maintenance was needed to remove solid material. This was carried out with the reactors operating as normal, meaning the settling legs required an effective isolation for the work to be carried out. Unfortunately the design only provided a single valve to do this, despite company standards requiring two isolation valves. The isolation procedure required that this single, actuated valve was secured in place by disconnecting its air supplies. Removing solids from the settling legs could be difficult and a 'work around' procedure [11] had been developed where the isolation valve would be operated to admit process pressure to dislodge the blockage. The investigation team identified that at the time of the incident, the control air supply to the actuated valve was reconnected in reverse (the air supply to the outlet and vice versa), which opened the valve rather than closing it (Fig. 4.1), resulting in loss of primary containment of flammable hydrocarbon from the reactor loop, which ignited and caused a vapour cloud explosion, killing 23 people and injuring over a hundred others. The investigation team identified contributing causes included inadequate permitting system, lock-out/tag-out process, and use of a non-fail safe valve.

Although the air supply to the valve was a contributing factor, the decision to carry out the maintenance whilst the system was operational and the lack of proper isolation were far more significant. This incident highlights that a single valve isolation should never be relied on for intrusive maintenance when there may be hazardous materials present, particularly under pressure. Isolations must be undertaken under a permit to work process, properly secured

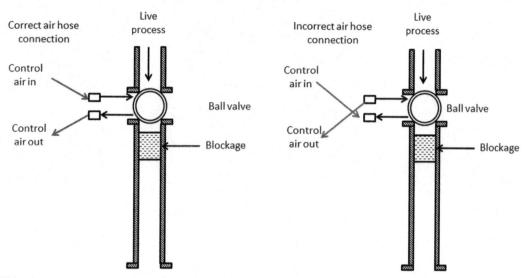

FIG. 4.1  Settling leg showing air hose connections. The valve would have stayed closed with the correct connections but it opened when the hoses were connected incorrectly [11].

(locked-out) to prevent any operation, and any return to service must be carefully controlled using a permit to work process, particularly when part of a system has been under maintenance whilst connected to systems that have been operating.

This explosion led to the setting up of one of the Mary Kay O'Connor Process Safety Center in Texas, set up by the husband of the operations superintendent Mary Kay O'Connor who was one of the 23 killed at the plant that day.

The fact that one of the world's foremost process safety centres was set up after a failure of activities associated with plant isolations and permit to work is a powerful reminder that catastrophic process safety events are not always due to failure of equipment associated with control of process (like relief devices and safety instrumented systems) and equipment integrity (like corrosion), but also activities associated with managing maintenance risk (like isolations and plant reinstatement) and failure in the original design process to provide equipment allowing safe isolation.

## 4.3.2  Isolating systems

Isolation prevents hazardous materials and energy from entering a system being maintained. Unfortunately, achieving an effective isolation is not always as simple as it may appear.

### 4.3.2.1  Process isolation

We have two main choices when isolating a process system for maintenance:

1. Positive isolation using a blank or blind flange (Kletz tended to call these 'slip plates') or by removing a section of pipe (Fig. 4.2);
2. Closing one or more valves.

**FIG. 4.2**   Devices available to create positive isolation [12]. *Contains public sector information published by the Health and Safety Executive and licensed under the Open Government Licence.*

Kletz provided a number of examples to illustrate why positive isolation should always be our first consideration. You will notice the date of this first incident, which may cause us to reflect how long we have been learning and relearning the same lessons [7].

---

**Text Box 4.9**

In 1928 a 36 in. (0.91 m) diameter low pressure gas line was being modified and a number of joints had been broken. Before work started the line was isolated by a closed isolation valve from a gasholder containing hydrogen, swept out with nitrogen and tested to confirm that no flammable gas was present. Unknown to the men on the job, the isolation valve was leaking. Eight hours after the job started the leaking gas ignited, there was a loud explosion and flames appeared at a number of the flanged joints on the line (Fig. 2.1). One man was killed, by the pressure, not the flames, but damage was slight.

The source of ignition was a match, struck by one of the workmen near an open end so that he could see what he was doing. He thought it was safe to strike a match as he had been assured that all flammable gas had been removed from the plant. Once a flammable mixture is formed a source of ignition is always liable to turn up, so the real cause of the explosion was not the match but the leaking valve. The match was merely the triggering event.

The report on the accident made three recommendations:

**(1)** Never trust an open gas main which is attached to a system containing gas, and keep all naked lights clear.
**(2)** When working on pipebridges at night, adequate lighting should be avail-able.
**(3)** Never place absolute reliance on a gasholder valve, or any other valve for that matter. A slip-plate is easy to insert and absolutely reliable.

The third recommendation was repeated in a Safety Handbook given to every employee the following year.

---

Here is another example from Kletz [13].

---

**Text Box 4.10**

A reactor was prepared for maintenance and washed out. No welding needed to be done, and no entry was required, so it was decided not to slip-plate off the reactor but to rely on valve isolations. Some flammable vapour leaked through the closed valves into the reactor and was ignited by a high-speed abrasive wheel which was being used to cut through one of the pipelines attached to the vessel. The reactor head was blown off and killed two men. It was estimated that 7 kg of hydrocarbon vapour could have caused the explosion.

After the accident, demonstration cuts were made in the workshop. It was found that as the abrasive wheel broke through the pipe wall a small flame occurred, and pipe itself glowed dull red.

The explosion could have been prevented by isolating the reactor by slip-plates or physical disconnection. This incident shows that valves are not good enough.

Isolation of service lines

A mechanic was affected by fumes while working on a steam drum. One of the steam lines from the drum was used for stripping a process column operating at 30 psi (2 bar). A valve on the line to the column was closed, but the line was not slip-plated. When the steam pressure was blown off, vapours from the column came back through the leaking valve into the steam lines.

The company concerned normally used slip-plates to isolate equipment under repair. On this occasion no slip-plate was fitted because it was a steam line. However, steam and other service lines in plant areas are easily contaminated by process materials, especially when there is a direct connection to process equipment. In these cases, the equipment under repair should be positively isolated by slip-plating or disconnection before maintenance.

When a plant was taken out of use, the cooling water lines were left full of water. Dismantling started nearly 20 years later. When a mechanic cut a cooling water line open with a torch, there was a small fire. Bacteria had degraded impurities in the water, forming hydrogen and methane.

---

However, it is far too simplistic to say that positive isolation is always required. Creating a positive isolation is intrusive maintenance in its own right. It involves breaking of one or more joints, which itself will normally be carried out with a valve isolation in place. Although positive isolation is inherently the safest solution for the maintenance task itself, it inevitably involves break of containment and so may not be the lowest risk option when all factors are taken into account. In practice most process isolations are achieved with valves, which is acceptable as long as the risks are understood and carefully controlled.

An incident occurred at a refining plant's sulphuric acid alkylation unit in 2015 whilst preparing a section of plant for maintenance [5]. One of the isolation valves normally used as an isolation point was known to be passing so the extent of the isolation was expanded (i.e. other valves were closed). Unbeknown to the operator, one of the new isolation valves was also passing. This allowed flammable process fluid to back-flow into a section that was assumed to contain only non-flammable fluids. As draining continued this flammable fluid entered the oily water system and then the sewer where it ignited. An operator received serious burns as a result. This incident highlights that isolations must be robust and integrity must be confirmed and not assumed.

Guidance has been published in the UK [12] explaining when valve isolation is acceptable and what form it should take. The guidance specifies a number of requirements when designing an isolation including:

- Each part of the isolation should be proved separately (e.g. prove each valve in a double block and bleed scheme);
- Each part should be proved to the highest pressure which can be expected within the system during the work activity. Particular care is required when there is a low differential pressure across valves where the sealing mechanism is activated by pressure.
- Where possible, each part of the isolation should be proved in the direction of the expected pressure differential.

This guidance has been available for many years and many companies and individuals have accepted it to be relevant and correct. However, evidence suggests that the requirements are often not followed in practice. This creates a large disconnect between theory and practice, which could result in risks being underestimated and hence improperly controlled. The solution is not simple, but being open about when the guidance cannot be followed will at least ensure that risks are better understood and controlled [14].

For hazardous systems under pressure the default isolation is usually described as 'double block and bleed' because it consists of two isolation valves in series with a vent or drain in between that can be used to bleed off pressure from the inter-space. It appears to provide an inherently reliable solution to ensure there is zero energy in the system before breaking containment, but in practice there can be issues. It must be recognised that valves can fail and multiple failures mean it cannot be assumed that closing two valves will necessarily result in an effective isolation. It is important that the integrity is confirmed for all valves forming the isolation and use of double block and bleed increases the complexity of doing this and hence the likelihood of human error. Double block and bleed is not necessarily the safest method of valve isolation and in some cases (e.g. low pressure liquid systems) a single valve isolation may create less risk. A balance has to be struck so that reducing the risk to personnel carrying out maintenance does not result in a higher risk to the people implementing the isolation [15].

Marking isolations on a process and instrumentation diagram (P&ID) helps prevent the wrong valve being isolated. It is not adequate to describe the valve to be isolated as the "one downstream of a the let-down pressure control valve," because there could be more than one, and when you are in the field, it's very easy to choose the wrong valve, which has in the past resulted in a wrong isolation, and overpressure of a line not rated to the upstream pressure.

### 4.3.2.2 Electrical isolation

Electrical isolation can appear to be a simple requirement but as Kletz highlighted people can be fooled quite easily into thinking they are safe when they are not [13].

---

### Text Box 4.11

When an electrical supply has been isolated, it is normal practice to check that the right switches have been locked or fuses removed before trying to start the equipment that has been isolated. However, this system is not fool proof.

In one case, the wrong circuit was isolated, but the circuit that should have been isolated was dead because the power supply had failed. It was restored while work was being carried out. In another case, the circuit that should have been isolated fed outside lighting. The circuit was dead because it was controlled by photo-eye control.

Electrical isolation points are usually remote from the systems they power. Kletz illustrated how an isolation can sometimes affect more plant than is immediately obvious, especially if drawings are not up to date [13].

### Text Box 4.12

On several occasions, maintenance teams have not realised that by isolating a circuit they have also isolated equipment that was still needed. In one case, they isolated heat tracing tape and, without realising it, also isolated a ventilation fan. The wiring was not in accordance with the drawings. In another case, maintenance team members isolated a power supply without realising that they were also isolating the power to nitrogen blanketing equipment and an oxygen analyser and alarm. Air leaked into the unit and was not detected, and an explosion occurred.

#### 4.3.2.3 Securing isolations

Isolations are only a temporary arrangement and so can be undone relatively easily. They need to be secured to prevent anyone else removing them (e.g. opening a valve or energising an electrical circuit) either because they do not know about the isolation or because they operate the wrong valve or switch in error.

Physical locks with unique keys can be the best method of securing an isolation. It is usually expected for electrical isolations and switchgear is usually designed with this in mind. This is not the case for process valves and some cannot physically be locked in position.

Locks can always be removed and so it is not only their physical characteristics that are important but the fact that they indicate an isolation is in place. Written tags may be considered an alternative provided everyone working on the site understands that they never touch an item with an isolation tag attached without formal permission. Multi-hasp locks can be used where more than one job or trade is relying on the same isolation. If more than one lock is used, this reduces the risk that when one job is finished, the isolation is removed before all jobs and trades have finished relying on that common isolation.

Actuated valves can be used as part of an isolation but often cannot be physically locked in position. In these cases the motive power (pneumatic, hydraulic of electrical) should be isolated and ideally physically disconnected. Trapped pressure should be vented off. Valves that fail open when their motive power is removed must never be used as a closed isolation, although they may still form part of an isolation if it requires the valve to be open (e.g. to create a vent).

Many companies refer to the method of securing isolations as Lock Out Tag Out (LOTO). It is a useful mnemonic to help people understand what is required.

#### 4.3.2.4 Self-isolation

Kletz shared his concerns about the practice where someone performing maintenance will implement the required isolations themselves [8].

**Text Box 4.13**

Preparation and maintenance by same person

A crane operator saw sparks coining from the crane structure. He turned off the circuit breaker and reported the incident to his supervisor who sent for an electrician. The electrician found that a cable had been damaged by a support bracket. He asked a crane maintenance worker to make the necessary repairs and reminded him to lock out the circuit breaker. The maintenance worker, not realising that the breaker was already in the 'off' position, turned it to the 'on' position before hanging his tag on it. While working on the crane he received second and third degree burns on his right hand.

The report said that the root cause was personnel error, inattention to detail, failure to confirm that the circuit was de-energised. However, this was an immediate cause, not a root cause. The root cause was a poor method of working.

Incidents are liable to occur when the same person prepares the equipment and carries out the maintenance. It would have been far better if the electrician had isolated the circuit breaker, checked that the circuit was dead and signed a permit-to-work to confirm that he had done so. It is not clear from the report whether the circuit was locked out or just tagged out. It should have been locked out and it should have been impossible to lock it open.

---

Self-isolation is used at many companies but should only be considered for specific low-risk situations that are clearly defined in company rules and procedures. Examples where it often occurs include process operators opening a filter to clean its element and instrument technicians isolating impulse pipework when calibrating a transmitter. One of the rules is usually that the workplace cannot be left with a self-isolation in place and, if the person does need to leave, the isolation is verified by someone else and tagged and recorded following 'normal' isolation procedures.

These self-isolations like filter change outs and instrumentation isolations are often formalised under what is known as a risk assessed procedure, and as such are undertaken outside the usual permit to work process. This can be appropriate because it allows the permit to work process to focus on the more hazardous activities.

### 4.3.3 Removing hazards

Some hazard will often remain after a system has been shutdown and so additional actions are required to remove this. Also, confirming system status, including whether it is free of hazard, should always be one of the last steps before maintenance starts.

The requirement to remove hazards needs to be considered when developing shutdown and isolation procedures. Particular problems can occur if items are isolated too soon because this can prevent hazards being removed effectively and establishing an isolation too late can remove the opportunity to prove its integrity (i.e. if pressure has been vented a valve cannot be checked for leaks).

Removing hazards effectively is not always as simple as it appears.

A deflagration followed by detonation occurred in a 6 in. ethylene pipeline whilst maintenance was being carried out injuring one of the two people present at the time [16]. The line had been pigged and flushed with nitrogen a few days before the incident. Valves were then closed to complete the isolation. The investigation concluded that the sequence of steps used to clear the piping had been inadequate with only a single valve separating the cleared pipework from a pressurised section of the pipeline. This valve leaked allowing ethylene to contaminate piping sections that had already been cleared. The pipeline design did not allow sampling to take place that was representative of the composition inside the pipeline. Ignition was from grinding taking place as part of the maintenance work.

Isolations must be robust and secured to prevent contamination of sections already cleared. If gas testing or other sampling is to be relied upon to confirm the hazard has been removed it must be truly representative of the whole system.

### 4.3.3.1 Cooling

Cooling of processes will normally commence as part of the shutdown so that energy can be removed in a controlled way. Whilst it can appear that cooling can only make things safe it must be recognised that it can cause problems. Examples include thermal strains if equipment is cooled too quickly. Also, allowing vapours to condense can cause vacuum, which can damage equipment that is not rated for negative pressures and leave liquids being trapped in the equipment that may cause a hazard during maintenance [3].

### 4.3.3.2 *Pumping out and draining*

Removal of liquid from a system requires careful consideration. Even water remaining in equipment is an issue because it can cause significant damage if it freezes or if it is heated and vaporises. Vulnerable items include control valves, pumps, heat exchanger and furnace tubes, cooling jackets, and steam systems. This can apply to spared items that can be easily overlooked when preparing systems for maintenance [3].

As much liquid as possible should be pumped out to a prescribed place where it can be handled appropriately before final draining takes place [3]. Care must be taken when levels get low so that pumps are not damaged by running dry. Gas may need to be admitted to assist final draining to below the level where pumps can operate, which is sometimes known as 'blowing clear'.

Hazardous liquids should be routed to closed systems to prevent people being exposed and avoid fires. Inert gases may need to be supplied during pumping and draining to prevent air ingress that can result in flammable atmospheres being formed.

Kletz highlighted that we need to be aware of where materials are going when removed from plant being prepared for maintenance [13].

---

### Text Box 4.14

A number of incidents have occurred because gas or vapour came out of drains or vents while work was in progress:

(a) Welding had to be carried out on a pipeline 6 m (20 ft) above the ground. Tests inside and near the pipeline were negative, so a work permit was issued. A piece of hot welding slag bounced off a pipeline and fell onto a sump 6 m below and 2.5 m (8 ft) to the side. The cover on the sump was loose, and some oil inside caught fire. Welding jobs should be boxed in with fire-resistant sheets.

Nevertheless, some sparks or pieces of slag may reach the ground. So drains and sumps should be covered.

**(b)** While an electrician was installing a new light on the outside wall of a building, he was affected by fumes coming out of a ventilation duct 0.6m (2ft) away. When the job was planned, the electrical hazards were considered and also the hazards of working on ladders. But it did not occur to anyone that harmful or unpleasant fumes might come out of the duct. Yet ventilation systems are installed to get rid of fumes.

And this was one of Kletz's scenarios in his Adventures of Joe series (Fig. 4.3) [8].

It is rarely possible to pump all liquid out of a system so some draining is often required. This can be conducted quite safely if systems are connected to a closed drain systems but this is not always possible. Also, connections to closed drain systems become another point that needs to be isolated because hazardous materials can back-flow into the system being maintained. Draining to open systems is often required at some point. Kletz highlighted how this can cause issues, especially if the drain points are high up [13].

**Text Box 4.15**

When a line is drained or blown clear, liquid may be left in low-lying sections and run out when the line is broken. This is particularly hazardous if overhead lines have to be broken. Liquid splashes down onto the ground. Funnels and hoses should be used to catch spillages. When possible, drain points in a pipeline should be fitted at low points, and slip-plates should be fitted at high points.

**FIG. 4.3**   Adventures of Joe—hazards when draining [8].

**FIG. 4.4**  Adventures of John—why a spring loaded drain valve can help [8].

In his Adventures of John, Kletz illustrated other problems with draining (Fig. 4.4) [8].

Draining vessels can appear to be a very simple activity but it can be hazardous. The Feyzin accident (see Appendix) that happened in France in 1966 killing 18 people and injuring 81 occurred when water was being drained from the bottom of a vessel containing Liquefied Petroleum Gas (LPG). The drain valves froze open when the water layer had been removed and LPG started to flow (see Section 1 of Appendix).

In 2011 an explosion occurred at the Chevron Refinery in Pembroke (UK) when a vessel was being prepared for maintenance (see Section 7 of Appendix). Four people were killed. One of the causes was a failure to drain the tank fully. The Operator relied on a level transmitter that was located 0.59 m above the base of the tank and so showed zero although liquid remained. Also, they did not understand the drain sump (see Fig. 4.5) that meant a cessation of flow from the drain was not a reliable indication that the tank was actually empty. The explosion occurred whilst residual liquid was being removed using a vacuum truck. The people involved in the task did not appreciate how much flammable liquid was present and so failed to take appropriate precautions [17].

### 4.3.3.3 Flushing

In some cases flushing with water or another fluid is required after draining to remove residual hazards. The choice of fluid and flushing conditions needs to be carefully considered and the contaminated flushing material has to be disposed of correctly.

In 2009 an explosion occurred in a sulphuric acid tank that was being prepared for maintenance [18]. Ignition was caused by grinding taking place to remove seized bolts so that a manway could be removed. Three people were injured. The investigation found that the contents of the tank had been drained and water had been used to flush out residual liquid.

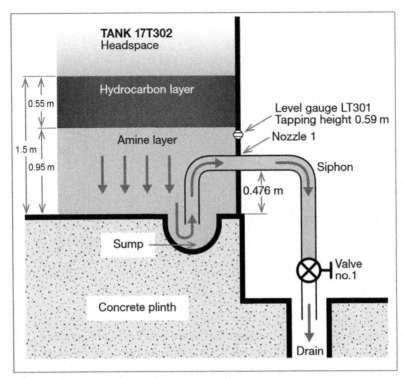

**FIG. 4.5** Schematic of the tank draining arrangement—Chevron Refinery Explosion [17]. *Contains public sector information published by the Health and Safety Executive and licensed under the Open Government Licence.*

Insufficient water was used and so a weak solution of sulphuric acid remained. This reacted with the tank's steel, which released hydrogen creating an explosive mixture inside the tank.

One of the reasons for flushing can be the potential for pyrophoric material to be encountered, which can ignite spontaneously when exposed to air, even at low temperatures [3]. Whilst some process materials can be pyrophoric, a common source is iron sulphide which can form when steel equipment is in contact with materials that contain sulphur.

In 2018 an explosion occurred in a Fluid Catalytic Cracking Unit (FCCU) at the Husky Superior Refinery, Superior, Wisconsin (USA) [19]. Thirty-six people sought medical attention as a result. It occurred whilst the FCCU was being shutdown for maintenance. The investigation concluded that air flowed from the unit's regenerator into downstream equipment. A slide valve had been closed to avoid this but it was leaking. Iron sulphide is a known issue in FCCU equipment. The shutdown plan included steps to mitigate this hazard, but that stage in the procedure had not yet been reached. The air mixed with hydrocarbon to form a flammable atmosphere and also reacted with the pyrophoric iron sulphide to cause ignition. Two vessels were destroyed by the explosion in which debris flew over 60 m (200 ft), puncturing a tank containing hot asphalt. Injuries to people were probably minimised because this occurred during a break time.

#### 4.3.3.4 *Pigging*

Preparing pipelines for maintenance can include pigging to remove hazards. It involves introducing a device known as a pig and propelling it through the pipeline using pressurised fluid. Pigging is used to displace liquids or solids (e.g. wax) that have settled on the bottom or stuck to the sides of the pipeline (internally).

Pigging can be used to carry out pipeline inspections. Like any other inspection, the data generated can be used to determine the condition of the pipeline and identify if maintenance is required.

All pigging operations are hazardous. They take place with the pipeline in an operational state and involve opening up systems to insert the pig at one end and remove it at the other, along with any fluids that may have been displaced. Pig launchers and receivers are designed to allow this to happen.

In 2013 an employee was killed when performing a routine pigging operation [20]. The pig had become stuck. It came loose whilst the receiver door was open, striking the employee. Wintery conditions had increased the likelihood of ice or hydrate formation, which was probably the reason the pig had got stuck.

Preparing a pig launcher or receiver for opening usually involves multiple isolations. Proving integrity of these isolations can require some complex procedures to be followed [14]. Use of interlocks that prevent steps being performed out of sequence is considered to be a standard requirement but it can add significant complexity and introduces other issues that can have a negative overall effect on the risks involved [21].

#### 4.3.3.5 *Clearing blockages*

Blockages can be the reason why maintenance is required in the first place and they can occur when a system is being prepared for maintenance, particularly if materials are allowed to cool. Kletz highlighted problems with the way blockages are sometimes cleared [8].

---

### Text Box 4.16

Clearing choked lines

Several incidents have occurred because people did not appreciate the power of gases or liquids under pressure and used them to clear choked lines.

For example, high pressure wash equipment was being used to clear a choked line. Part of the line was cleared successfully, but one section remained choked so the operators decided to connect the high pressure water directly to the pipe. As the pressure was 100 bar (it can get as high as 650 bar) and as the pipe was designed for only about 10 bar, it is not surprising that two joints blew. Instead of suspecting that something might be wrong operators had the joints remade and tried again. This time a valve broke.

Everyone should know the safe working pressure of their equipment and should never connect up a source of higher pressure without proper authorisation by a professional engineer who should first check that the relief system is adequate.

On another occasion gas at a gauge pressure of 3 bar—which does not seem very high—was used to clear a choke in a 2 in. pipeline. The plug of solid was moved along with such force that when it hit a slip-plate it made it concave. Calculation, neglecting friction, showed that if the plug weighed 0.5 kg and it moved 15 m, then its exit velocity would be 500 km/h!

An instrument mechanic was trying to free, with compressed air, a sphere which was stuck inside the pig chamber of a meter prover. Instead of securing the chamber door properly he fixed it by inserting a metal rod—a wheel dog—into the top lugs. When the gauge pressure reached 7bar the door flew off and the sphere travelled 230m before coming to rest, hitting various objects on the way.

In all these cases it is clear that the people concerned had no idea of the power of liquids and gases under pressure. Many operators find it hard to believe that a 'puff of air' can damage steel equipment.

### 4.3.3.6 Purging

Purging will usually involve displacing gases and vapours with inert gases such as nitrogen or steam. The aim is to reduce risks by replacing a hazardous substance with a less hazardous one.

Kletz talked quite a lot about purging in his books, with this incident illustrating why it is important [13].

### Text Box 4.17

A bottom manhole was removed from an empty tank still full of gasoline vapour. Vapour came out of the manhole and caught fire. As the vapour burned, air was sucked into the tank through the vent until the contents became explosive. The tank then blew up.

One of the most common purges is to displace flammable gases with an inert gas, typically nitrogen or steam. This can be completed as a flow purge, where the inert gas enters at one location and is exhausted at another, or, alternatively, as a pressure purge, where the system is pressurised with inert gas, given time to mix, and then released.

Purging continues until the concentration of the toxic or flammable gas or vapour is considered safe. This may be confirmed by carrying out a gas test.

One of the challenges is that vapours can be removed successfully but there are ways that they can return. This can be the result of a poor isolation but sometimes it is because of residual liquids that vaporise. Kletz illustrated this with his Adventures of Joe (Fig. 4.6) [8].

Kletz highlighted that similar issues can result with solids [13].

### Text Box 4.18

Solids in a vessel can 'hold' gas that is released only slowly. A reactor, which contained propylene and a layer of polypropylene granules 1 to 1.5m thick, had to be prepared for maintenance. It was purged with nitrogen six times. A test near the manhole showed that only a trace of propylene was present, less than 5% of the lower explosive limit (LEL). However, when the reactor was filled with water, gas was emitted, and gas detectors in the surrounding area registered 60% of the LEL.

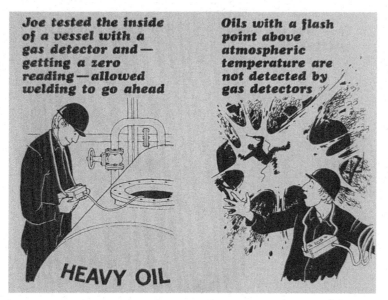

FIG. 4.6    Adventures of Joe—gas testing is not fool proof [8].

Kletz showed that empty or even new tanks can be hazardous [13].

---

### Text Box 4.19

A label had to be welded onto an empty drum. As the drum was brand new, no precautions were taken, and no tests were carried out. The drum exploded, breaking the welder's leg. The manufacturer had cleaned the drum with a flammable solvent, had not gas freed it, and had not warned the customer.

---

A trap that people sometimes fall into is believing that, after a system has been purged and proven safe by a gas test, it will remain safe for ever more. Kletz illustrated this with the Adventures of Joe (Fig. 4.7) [8].

In the following example Kletz highlighted that hazards can remain undetected for a very long time but show themselves during maintenance [13].

---

### Text Box 4.20

An old propylene line that had been out of use for 12 years had to be modified for reuse. For the past 2 years it had been open at one end and blanked at the other. The first job was welding a flange onto the open end. This was done without incident. The second job was to fit a 1-in. branch 60 m from the open end. A hole was drilled in the pipe and the inside of the line tested. No gas was detected. Fortunately, a few hours later, just before welding was about to start, the inside of the pipe was tested again, and flammable gas was detected. It is believed that some gas had remained in the line for 12 years and a slight rise in temperature had caused it to move along the pipeline.

---

**FIG. 4.7** Adventures of Joe—hazards can return after equipment has been prepared for maintenance [8].

### 4.3.3.7 Preparing confined spaces for people to enter

Definitions of 'confined space' vary around the world but they are enclosed places where the working environment can be hazardous to people because of the potential for a build-up of hazardous materials creating a toxic, flammable or low oxygen atmospheres, or the potential creation of hazardous conditions by other energy sources such as steam or mechanical movement. Some confined spaces, such as inside process tanks and vessels, are obvious. Others are not.

The aim must always be to avoid the need for anyone to enter a confined space. The use of drones and other remotely operated devices is starting to provide practical alternatives; however, they will not cover every situation. Whenever someone is required to enter a confined space it is essential that it is positively isolated from any sources of hazard. Pipework should be disconnected wherever possible in preference to inserting slip plates, spades, or blinds, to ensure that the status is visible. Additional efforts must be made to remove any hazardous materials and to ensure a safe atmosphere is present whenever someone is inside. Use of breathing apparatus may be required but should only be considered after all efforts have been made to make the space safe. Arrangements for rescue should also be confirmed as in place before entry is commenced.

Kletz identified many accidents involving entry to confined spaces, including [13]:

- Vessel not fully freed from hazardous material because of a physical divide (baffle) meaning only one end could be cleaned and checked before entry;
- Person overcome by fumes from the residue of liquid used to wash out the normal contents of the tank;
- Oxygen deficient atmosphere in a water tank caused by rust formation;
- Flammable or toxic liquids trapped in the bearings of stirrers that then leak out;

- People overcome by solvent taken into a tank for carrying out a dye-penetration test;
- Flammable atmosphere when painting inside a confined space;
- Vessels not positively isolated from danger;
- People harmed when steam lines running through a confined space failed and people could not escape quickly;
- People inside a vessel harmed when internal machinery was started;
- Nitrogen entering the space because of a leaking valve (all nitrogen supplies should be disconnected before entry);
- People unfamiliar with the hazard entering confined space without authorisation.

Kletz highlighted that people are often harmed when attempting to rescue another person overcome inside a confined space. He described this as 'misguided bravery' that occurs because of a strong natural impulse to wade in and rescue the individual.

One of the steps in preparing a vessel or tank for entry (purging) is a common cause of accidents. Although the purpose of purging is to remove hazards, inert gases are also hazardous. Whilst purging reduces the risk of fire, explosion, and toxic exposure the potential for people to be harmed by the purge medium must be recognised and managed. Kletz said that nitrogen 'has killed more people than any other substance' (he also said that water was probably our most dangerous liquid) [22]. He also said that calling purge gases including nitrogen 'inert' can be misleading because people can understand this to mean it is harmless, which is definitely not the case.

One person died from nitrogen asphyxiation when a tarpaulin was draped over an open manway on a column that was being purged with nitrogen [3]. Their colleague survived because he collapsed in a place where he received sufficient air to sustain life. They were there to carry out inspections of the flange face and the investigation concluded that they had arranged the tarpaulin to provide shelter from the wind or to block light so that they could see the dye penetrant that they were using. Whatever the reason it was sufficient to create a hazardous confined space.

Kletz used the Adventures of Joe series to illustrate the hazards of inert gases that are often used for purging (Fig. 4.8) [8].

If people are required to enter a vessel or even work nearby it is essential to ventilate it after purging, to displace the inert gas with fresh air.

### 4.3.3.8 Gas testing

Gas testing is frequently carried out to confirm the atmosphere inside a vessel or other item is safe. This can be to confirm flammable materials have been removed before air is introduced; or that toxic or inert materials have been removed before people enter.

People do not always appreciate how to perform gas tests correctly. The availability of sophisticated electronic devices can give the impression that they only need to know how to use the equipment. But this is far from the truth and failure to understand the purpose of the test and misinterpreting the results can have very serious consequences. Unfortunately, human behaviour means that once someone believes something is safe (i.e. because the gas test says it is) they tend to relax their attention to potential hazards, believing controls that may be considered normal in other circumstances no longer apply.

In the 2011 explosion at the Chevron Refinery in Pembroke (UK) (see Section 7 of Appendix) a gas test was carried out before using a vacuum truck to remove residual liquid from a

**FIG. 4.8**   Adventures of Joe—inert gases may remove some hazards but they are harmful to people [8].

tank. The test was taken from the top of the tank and so could not be representative of the atmosphere in the tank. The result obtained was 67% of the Lower Explosion Limit (LEL). This should have rung alarm bells because normally a value of 10% of LEL is considered to be the safe limit (occasionally 25% of LEL is sometimes permissible for certain activities). Unfortunately, the implications of the gas test result were not properly considered and a static spark or exposure of pyrophoric material caused ignition of hydrocarbon/air mixture inside the tank [17].

### 4.3.3.9 Removing stored energy

Process systems can store energy as pressurised gas and vapours. These should have been removed by actions take to remove hazards. However, pockets can remain in unexpected places.

Kletz shared a number of examples where pressure was released during maintenance causing a hazard [13].

---

**Text Box 4.21**

Even though equipment is isolated by slip-plates and the pressure has been blown off through valves or by cracking a joint, pressure may still be trapped elsewhere in the equipment, as the following incidents show:

**(a)** This incident occurred on an all-welded line. The valves were welded in. To clear a choke, a fitter removed the bonnet and inside of a valve. He saw that the seat was choked with solid and started to chip it away. As he did so, a jet of corrosive chemical came out under pressure from behind the solid, hit him in the face, pushed his goggles aside, and entered his eye.

**(b)** An old acid line was being dismantled. The first joint was opened without trouble. But when the second joint was opened, acid came out under pressure and splashed the fitter and his assistant in their faces. Acid had attacked the pipe, building up gas pressure in some parts and blocking it with sludge in others.

**(c)** A joint on an acid line, known to be choked, was carefully broken, but only a trickle of acid came out. More bolts were removed, and the joint pulled apart, but no more acid came. When the last bolt was removed and the joint pulled wide apart, a sudden burst of pressure blew acid into the fitter's face.

In all three cases the lines were correctly isolated from operating equipment. Work permits specified that goggles should be worn and stated, "Beware of trapped pressure".

To avoid injuries of this sort, we should use protective hoods or helmets when breaking joints on lines that might contain corrosive liquids trapped under pressure, either because the pressure cannot be blown off through a valve or because lines may contain solid deposits.

In 2001 an incident occurred when a cover on a vessel blew off whilst being removed at BP Amoco's polymer site in Augusta, Georgia (USA) [23]. Hot molten polymer had decomposed in the vessel when production stopped due to problems downstream. At first the vessel filled with a foam that hardened around the inside walls of the vessel. But the reaction continued closer to the centre of the vessel creating gas that pressurised the vessel. However, the hardened polymer had blocked a pressure gauge on the vessel so that personnel could not determine the pressure. Believing that the vessel was no longer under pressure they started to remove the cover. The energetic release damaged nearby pipework, releasing flammable material. A fire occurred and three people died as a result.

Other sources of stored energy include pneumatic and hydraulic systems (including facilities providing motive force for valves), electrical equipment (batteries and capacitors), and mechanical items (springs and suspended loads). This is another factor to consider when planning isolations to ensure that energy is not trapped between isolated points. Great care should always be taken when breaking into systems until there is clear evidence that no energy could be stored.

### 4.3.3.10 Removing dust

If dust is ever allowed to settle it can be disturbed by maintenance creating an explosion risk. Removal of dust layers is often included as a 'housekeeping' activity. This label may trivialise its importance. Removing dust build-ups should be considered as essential maintenance.

In March 2011 an engineer was severely burned at the Hoeganaes Gallatin facility in Tennessee (USA). He was using a metallic hammer during replacement of igniters on a band furnace. The work resulted in a large amount of combustible dust being lifted from flat surfaces of the furnace. The investigation concluded that the use of the hammer on metallic surfaces created the spark that caused ignition and that preventing dust formation and removing it before maintenance could have prevented this accident. Also, a rubber or wooden mallet should have been used because it would not have created a spark.

## 4.4 Safety during maintenance

Whilst a lot of the risk associated with maintenance can be eliminated or significantly reduced by the steps taken to prepare systems, there is still plenty of opportunity for accidents during maintenance. Many of these are the more 'normal' occupational health and safety, which is not really the focus of this book, but process safety and major accidents remain very relevant.

### 4.4.1 Is it safe to proceed?

There are two main concerns when deciding if it is safe for maintenance work to take place:

1. Has the system been prepared correctly?
2. Are the methods to be used during the maintenance safe?

In 2019 two rail workers were killed by a train [24]. Paperwork analysed after the accident indicated that it would have been safe to carry out the work later in the day (after 12:30) because that was then the track was going to be closed to traffic. Unfortunately the team started work much sooner than this (at 08:50). It is not clear whether there was confusion around the time when the work could start or whether workers thought they could do their job safely on a live track. It was reported that there was a "general lack of understanding" of how paperwork should be interpreted when deciding when to start work.

Kletz highlighted that a permit to work system is usually used to control when maintenance work can proceed but as a procedural control it has its limitations due to human factors [8].

---

### Text Box 4.22

Incidents have occurred because permit to work procedures were not followed. Often it is operating teams who are at fault but sometimes the maintenance teams are responsible.

Common faults are:

- Carrying out a simple job without a permit-to-work;
- Carrying out work beyond that authorised on the permit-to-work;
- Not wearing the correct protective clothing.
  To prevent these incidents we should:
- Train people in the reasons for the permit system and the sort of accidents that occur if it is not followed. There is, unfortunately, no shortage of examples
- Check from time to time that the procedures are being followed.

---

#### 4.4.1.1 *Risk assessment and method statement*

It is well established that all work should be subject to risk assessment and as a result of this a method for doing the work can be defined. The acronym RAMS (Risk Assessment and Method Statement) has been widely adopted to cover this process when planning maintenance. Often it is a requirement to submit the RAMS for approval before work can commence, especially for infrequent jobs and those performed by contractors.

Although the principles are correct the use of RAMS is not always effective. One of the main shortcomings is that they only cover the work to be carried out by maintenance personnel, and do not take account of the status of the system to be worked on or the potential for other activities to be occurring at the same time. Also, RAMS are often viewed as a bureaucratic requirement meaning generic versions, often with little detail, are produced. The people supposed to be reviewing them are easily reassured because they receive a document and can 'tick the box' on their approval checklist. Once given permission to start the job no one actually refers to the RAMS because they contain no practical or useful information.

### 4.4.1.2 *Maintenance procedures and competence*

Although some progress has been made with defining when operating procedures are required and what they should cover this is not always the case for maintenance. Procedures should have a role in the management of maintenance risks but there are some issues to overcome.

One of the complaints made when maintenance procedures are discussed is that systems are made up of many different components and all may need to be maintained at some time. If a procedure was required for each maintenance task it would create a massive workload and would be unmanageable. Another argument is that most maintenance tasks are not specific to any single item of equipment and so people with the right skills can perform maintenance using their competence and so do not require any procedures.

At present detailed maintenance procedures are very rare. The requirement to perform a task is identified via a maintenance management system, which may include a high level description of the main activities that need to be performed. People considered competent are then expected to refer to available documents, including vendor manuals and drawings, to determine how to perform the task. RAMS may be produced, but are largely focussed on the personal safety aspects of the person doing the work and overlook most process safety concerns.

Reliance on competent people provided with basic information is probably acceptable for most simple, frequently performed maintenance tasks. It is probably insufficient for more complex and hazardous activities. Companies should have processes in place to determine when maintenance personnel require the support of documented procedures that take these factors into account. These procedures should be presented in the most useful format, which may include a checklist to ensure the appropriate ISO or ASME industry standard is followed. Using the procedures, when identified as necessary, should become an integral part of the work.

### 4.4.1.3 *Realistic controls*

Kletz highlighted that it is easy when planning maintenance from an office to specify impractical controls [8].

---

### Text Box 4.23

A refinery compressor was isolated for repair and swept out with nitrogen but, because some hydrogen sulphide might still be present, the fitters were told to wear air-line breathing apparatus. They found it difficult to remove a cylinder valve which was situated close to the floor, so one fitter decided to remove his mask and was overcome by the hydrogen sulphide. Following the incident lifting aids were provided.

Many companies would have been content to reprimand the fitter for breaking the rules. The company concerned, however, asked why he had removed his mask and it then became clear that he had been asked to carry out a task which was difficult to perform while wearing a mask.

Ensuring people planning work have a good understanding of what is involved is important to make sure they specify practical controls. If the planner has not done the job themselves they should get input from people who have, which may include contractors. Also, it is important that people doing the work understand that they do need to implement the specified controls and they should always stop work and seek approval before making any changes.

### 4.4.1.4 Contractors

A lot of maintenance is performed by contractors and other third parties including vendors. Kletz had concerns about this [8].

#### Text Box 4.24

Many accidents have occurred because contractors were not adequately trained.

For example, storage tanks are usually made with a weak seam roof, so that if the tank is overpressured the wall/roof seam will fail rather than the wall/floor seam. On occasions contractors have strengthened the wall/roof seam, not realising that it was supposed to be left weak.

Many pipe failures have occurred because contractors failed to follow the design in detail or to do well what was left to their discretion. The remedy lies in better inspection after construction, but is it also possible to give contractors' employees more training in the consequences of poor workmanship or short cuts on their part? Many of them do not realise the nature of the materials that will go through the completed pipelines and the fact that leaks may result in fires, explosions, poisoning or chemical burns. Many engineers are sceptical of the value of such training. The typical construction worker, they say, is not interested in such things. Nevertheless, it might perhaps be tried.

Whilst Kletz's concerns remain valid, contractors will continue to perform maintenance. It is important to understand how this results in risk due to issues such as lack of local knowledge and barriers to communication and coordination. Also, it is worth noting the benefits of using contractors and ensuring work is arranged to maximise these, including allowing contractors to use their expertise and taking their advice.

A serious derailment occurred at Potters Bar, UK in 2002 [25]. Seven people died and over 70 were injured. The direct cause was a mechanical failure of points. Maintenance of the track including these points had been given to a contractor. The investigation expressed the view that 'use of contractors for maintenance work does not, or itself, compromise the integrity of the rail infrastructure'. However, contracts need to provide the right incentives to ensure attention to safety, the client needs to be knowledgeable of the risks, and the complexity introduced by the arrangement has to be understood and managed.

### 4.4.1.5 Permit to work

Permit to work systems have become standard in many industries. They are intended to provide a mechanism to control maintenance work to reduce risks. Kletz described in very basic terms how the requirement to obtain a permit before starting work contributes to safety [7].

---

**Text Box 4.25**

Every day on every plant, equipment which has been under pressure is opened for repair, normally under a permit-to-work system. One man prepares the equipment and issues a permit to another man who opens up the equipment, usually by carefully slackening the bolts in case any pressure is left inside. The involvement of two people and the issue of a permit provides an opportunity to check that everything necessary has been done. Accidents are liable to happen when the same man prepares the equipment and opens it up.

---

A permit to work is usually a piece of paper that is issued as part of the process of starting a maintenance or other routine task. However, the overall system is far more than this.

The UK's Health and Safety Executive has published guidance that states "A permit-to-work system aims to ensure that proper consideration is given to the risks of a particular job or simultaneous activities at site" [26]. It provides a list of objectives and functions that can be summarised as follows:

- Formally authorising certain types of work;
- Clarifying what work is going to be carried out including the methods and tools to be used;
- Identifying the hazards that may be encountered due to the work to be carried out, the systems being worked on, and adjacent activities;
- Identifying preparations required before the work can start, including isolation;
- Identifying the precautions to be taken whilst the work is taking place;
- Ensuring the people usually in control of the items being worked on understand what is happening;
- Providing a procedure for suspending partly finished work;
- Coordinating different pieces of work to ensure they do not interfere with each other;
- Providing formal handover mechanisms between the people usually in control and the work party, including handback on completion;
- Ensuring all changes to work scope, or methods are properly controlled;
- Communicating the contents of the permit to people who may be effected;
- Defining actions to be taken if an emergency occurs;
- Monitoring and auditing work and its control.

This list highlights why a permit to work system needs to be well defined and managed by competent people. However, there is a danger if it becomes bureaucratic and time consuming. This was illustrated by Kletz [8].

---

**Text Box 4.26**

Many supervisors find permit-to-work procedures tedious. Their job, they feel, is to run the plant, not fill in forms. There is a temptation, for a quick job, not bother. The fitter is experienced, has done the job before, so let's just ask him to fix the pump again.

They do so and nothing goes wrong, so they do so again. Ultimately the fitter dismantles the wrong pump, or the right pump at the wrong time, and there is an accident. Or the fitter does not bother to wear the proper protective clothing requested on the permit-to-work. No-one says anything, to avoid unpleasantness; ultimately the fitter is injured.

To prevent these accidents, and most of those described in this chapter, a three-pronged approach is needed:

(1) We should try to convince people that the procedure - in this case a permit-to-work procedure - is necessary, preferably by describing accidents that have occurred because there was no adequate procedure or the procedure was not followed. Discussions are better than lectures or reports.

The discussion leader should outline the accident and then let the group question him to find out the rest of the facts. The group should then say what they think should be done to prevent the accident happening again.

(2) We should make the system as easy to use as possible and make sure that any equipment needed, such as locks and chains and slip-plates, is readily available. Similarly, protective clothing should be as comfortable to wear as possible and readily available. If the correct method of working is difficult or time-consuming then an unsafe method will be used.

(3) Managers should check from time to time that the correct procedures are being followed. A friendly word the first time someone takes a short cut is more effective than punishing someone after an accident has occurred.

If an accident is the result of taking a short cut, it is unlikely that it occurred the first time the short cut was taken. It is more likely that short-cutting has been going on for weeks or months. A good manager would have spotted it and stopped it. If he does not, then when the accident occurs he shares the responsibility for it, legally and morally, even though he is not on the site at the time.

As already stated, a manager is not, of course, expected to stand over his team at all times. But he should carry out periodic inspections to check that procedures are being followed and he should not turn a blind eye when he sees unsafe practices in use.

The first step down the road to a serious accident occurs when a manager turns a blind eye to a missing blind.

---

One of the reasons why permit to work systems are less effective than they should be is overuse. Insisting a permit is required for every maintenance task or every task performed by a contractor means a lot of time is spent generating paperwork that adds very little value.

Electronic permit to work systems are now available that can, in theory, reduce some of the bureaucratic burden. Care must be taken to ensure the chosen application supports the underlying processes and that it does not start to determine how work is controlled in practice.

Ultimately it must be recognised that a permit to work system is only effective if it supports the people involved in the process. It is interesting to note that the UK's Health and Safety Executive identify permit to work as a critical communication rather than a work control procedure [27].

### 4.4.1.6 Toolbox talks

It is easy for people planning maintenance from an office to assume that all the controls they have identified will be implemented and effective. In practice it is very easy for hazards to be overlooked or for the people performing work to misinterpret the requirements or fail to understand their importance, or the work on the day to be different to that planned due to changes not originally considered.

Short, informal safety meetings at the workplace or very close to it, just before the work activity commences, have proven to be effective at focussing attention on safety. They are often known as 'toolbox talks' (this term is used more generally for all types of short, informal meetings). They provide an excellent opportunity to carry out a last minute check of site status and condition, and to look for unexpected hazards. More importantly, they provide an opportunity for the people involved in planning and authorising work to discuss it with the people about to carry it out. Some companies specify that toolbox talks or similar take place at the worksite before every piece of work, and repeat them at the restart after breaks during the day.

A good pre-job toolbox talk goes beyond checking the work party understands how you could get hurt (what could crush you, how could you fall from height, how could an object be dropped, or stored energy released?), what controls keep us from getting hurt, and what mitigations should be in place in the event of an emergency. It seeks to question how the work could be more difficult today, so these issues can be discussed before work can start safely. It checks whether we have a full work party, or any new members, and enough time to do the task. It enquires about adjacent activities that may impact our work or the permit. It encourages us to improve our situational awareness (is there sufficient visibility, confusing displays, bad weather?). It asks whether we have the right tools and materials for the job, or are we considering using workarounds to get the work done? Is there any part of the task that's unclear or particularly complex that we should talk through? The crew leader should also be looking out for signs of fatigue or distraction or stress in their work party because this can also impact safety. And lastly, a good toolbox talk also reinforces what we should do if things change during the work activity, for example if we move from the work plan into problem solving – we stop the job and review the permit and risk assessment with the permit issuer. Such discussions are best undertaken immediately before the work starts, and after each restart (like lunch) so that changes in the environment can be assessed before work can start safely.

### 4.4.1.7 *Point of work risk assessments*

Point of work risk assessments are another attempt to ensure people carrying out work are fully aware of the hazards that they may encounter. They require the work party to assess the status of the items that they are going to work on and to consider the area as a whole to determine what else is happening and to identify any hazards. They are intended to address the issue that things can change between the time a piece of work is planned and the time it is carried out, and that things may change between the start and finish of a piece of work.

Formalised approaches have been developed for point of work risk assessment. Some are proprietary systems that consist of short checklists that should be completed before starting a piece of work and at the start of each day if the work continues over a long time. Training in what to consider and how to use the checklist is usually provided.

Kletz highlighted that one of the key questions when starting any maintenance work on a process system is understanding how it has been isolated and ensuring that isolation is still intact [8].

## Text Box 4.27

Many accidents have occurred because the operating team failed to isolate correctly equipment which was to be repaired. It is sometimes suggested that the maintenance workers should check the isolations and in some companies this is required.

In most companies however the responsibility lies clearly with the operating team. Any check that maintenance workers carry out is a bonus. Nevertheless, they should be encouraged to check. It is their lives that air at risk.

On one occasion a large hot oil pump was opened up and found to be full of oil. The ensuing fire killed three men and destroyed the plant. The suction valve on the pump had been left open and the drain valve closed.

The suction valve was chain-operated and afterwards the fitter recalled that earlier in the day, while working on the pump bearings, the chain had got in his way. He picked it up and, without thinking, hooked it over the projecting spindle of the open suction valve!

This incident also involved a change of intention. Originally only the bearings were to be worked on. Later the maintenance team decided that they would have to open up the pump. They told the process supervisor but he said that a new permit was not necessary.

It is truly worrying how many maintenance personnel will start working on an item without any understanding of what has been done to ensure their safety. There are so many potential human errors when people are planning and implementing isolations that they should be expected. Assuming an item is safe because it says so on a piece of paper is definitely not sensible.

### 4.4.1.8 Equipment identification

One of the greatest risks of maintenance is working on the wrong piece of equipment. Kletz illustrated the problem [8].

## Text Box 4.28

On many occasions someone has broken the wrong joint, changed the wrong valve or cut open the wrong pipeline. Sometimes the job has been shown to the fitter who has gone for his tools and then returned to the wrong joint or valve; sometimes a chalk mark has been washed off by rain or the fitter has been misled by a chalk mark left from an earlier job; sometimes the job has been described but the description has been misunderstood. Equipment which is to be maintained should be identified by a numbered tag (unless it is permanently labelled) and the tag number shown on the permit-to-work. Instructions such as 'The pump you repaired last week is giving trouble again' lead to accidents.

Sometimes the process team members have identified the equipment wrongly. To stop a steam leak a section of the steam line was bypassed by a hot tap and stopple. After the job was complete it was found that the steam leak continued. It was then found that the leak was actually from the condensate return line to the boiler which ran alongside the steam line. The condensate was hot and flashed into steam as it leaked. Several people then said that they had thought that the steam seemed rather well! This incident could be classified as a 'mind-set'. Having decided that the leak was coming from the steam main, everyone closed their minds to the possibility that it might be coming from somewhere else.

Whilst labelling equipment is essential to reduce the likelihood of people working on the wrong item, Kletz highlighted how illogical basic design introduces significant risk [13].

---

### Text Box 4.29

The need for clear, unambiguous labelling

(a) A row of pumps was labelled (Fig. 4.9). A mechanic was asked to repair No. 7. Not unreasonably, he assumed that No. 7 was the end one. He did not check the numbers. Hot oil came out of the pump when he dismantled it.

(b) There were four crystallisers in a plant, three old ones and one just installed. A man was asked to repair A. When he went onto the structure, he saw that two were labelled B and C but the other two were not labelled. He assumed that A was the old unlabelled crystalliser and started work on it. Actually, A was the new crystalliser. The original three were called B, C, and D. Crystalliser A was reserved for a possible future addition for which space was left (Fig. 4.10).

(c) The labels on two air coolers were arranged as shown in Fig. 4.11. The B label was on the side of the B cooler farthest away from the B fan and near the A fan. Not unreasonably, workers who were asked to overhaul the B fan assumed it was the one next to the B label and overhauled it. The power had not been isolated. But fortunately, the overhaul was nearly complete before someone started the fan.

---

**FIG. 4.9**   Numbering pumps like this leads to errors.

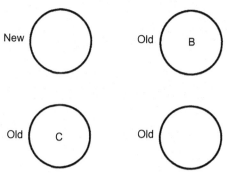

**FIG. 4.10**   Which is crystalliser A? [13].

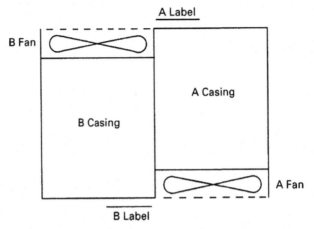

**FIG. 4.11**   Air coolers—which is the A fan? [13].

It is very unfortunate that helping people identify equipment is given very little consideration. Designers will often locate items arbitrarily and inconsistently, satisfying themselves that labelling will be totally effective for identification. Despite their importance labels are often poorly maintained. They fall off, become illegible, or are moved during maintenance and it seems no one has the task of checking and repairing them. There are sites that have very poor labelling, which adds complexity for people operating and maintaining systems, and significantly increases the risks of identification errors, which can have very serious consequences.

### 4.4.2 Carrying out maintenance safely

Most maintenance risks can be controlled through good planning and preparation. But there are still potential issues whilst maintenance is being carried out.

#### 4.4.2.1 *Dismantling equipment*

Most maintenance will involve dismantling equipment. This may be simply removing a cover for access or breaking joints in pipework or vessels. Kletz highlighted how errors when doing this can be hazardous [8].

---

### Text Box 4.30

**(a)** People have been injured when dismantling diaphragm valves because they did not realise that the valves can contain trapped liquid. Here again, a hardware solution is often possible - liquid will not be trapped if the valves are installed in a vertical section of line.

**(b)** On a number of occasions men have been asked to change a temperature measuring device and have removed the whole thermowell. One such incident, on a fuel oil line, caused a serious refinery fire.

**(c)** On several occasions, when asked to remove the actuator from a motorised valve, men have undone the wrong bolts and dismantled the valve. One fire which started in this way killed six men. On other occasions trapped mechanical energy, such as a spring under pressure, has been released.

A hardware solution is possible in these cases. Bolts which can safely be undone when the plant is up to pressure could be painted green; others could be painted reds. A similar suggestion is to use bolts with recessed heads and fill the heads with lead if the bolts should not be undone when the plant is up to pressure.

One of the most hazardous parts of any maintenance activity is the initial break of containment. A particular concern is working on the wrong item or starting work before it has been prepared properly. Also, hazards can remain even after preparation has been completed. Any break of containment needs to be carried out with great care. Kletz illustrated the potential for trapped pressure in his Adventures of Joe series (Fig. 4.12) [8].

### 4.4.2.2 Reassembly or reinstatement

The main risks during disassembly affect the personal safety of the people doing the work. However, there are far more opportunities for error during reassembly but the consequences are unlikely to be immediate. Incidents typically occur when returning the system to maintenance or plant reinstatement.

Using the wrong spare parts or consumable items such as gaskets can have catastrophic consequences if they fail with the system in operation. Kletz used the Adventures of John to illustrate these errors are easy to make but the likelihood can be reduced if spares policies and other systems are set up with this in mind (Fig. 4.13) [8].

More significant errors can occur. Kletz illustrated that it is sometimes not clear whether the available spare or replacement items are correct for the service [7].

FIG. 4.12   Adventures of Joe—stored energy is hazardous [8].

FIG. 4.13    Adventures of John—it is easy to choose the wrong gasket [8].

---

**Text Box 4.31**

During the night a shift fitter was asked to change a valve on a unit that handled a mixture of acids. He could not find a suitable valve in the workshop but after looking round he found one on another unit. He tested it with a magnet. It was non-magnetic so he assumed it was one of the valves made from grade 321 or 316 steel which were normally used and installed it. Four days later it had corroded and there was a leak of acid. The valve was actually made from Hastelloy, an alloy suitable for use on the unit where it was found but not suitable for use on the unit on which it was installed.

---

Making sure the correct spares are available before maintenance commences reduces the likelihood of people using the wrong ones. Collecting them together as a package for the task can prevent people having to go and find what they need from a general store. People carrying out maintenance still need to understand that they should not change any item without formal authorisation, but sometimes they are trapped into thinking that they are making a 'like for like' change and so there are no issues.

Kletz illustrated with his Adventures of John series that selecting the right parts is not enough if they are fitted incorrectly (Fig. 4.14) [8].

Kletz highlighted that many problems with reassembly can be avoided through good design [4].

FIG. 4.14    Adventures of John—some equipment has to be the right way around to work [8].

## Text Box 4.32

Friendly plants are designed so that incorrect assembly is difficult or impossible. For example, compressor valves should be designed so that inlet and exit valves cannot be interchanged.

With friendly equipment, it is possible to see at a glance if it has been assembled or installed incorrectly or whether it is in the open or shut position. For example, check (non-return valves) should he marked so that installation the wrong way round is obvious (it should not be necessary to look for a faint arrow hardly visible beneath the dirt), and gate valves with rising spindles are friendlier than valves with non-rising spindles because it is easy to see whether they are open or shut. Ball valves are friendly if the handles cannot be replaced in the wrong position.

However, apparently fool-proof designs are not always as reliable as they may appear during design. People are very able to defeat even the most inherent of safety features. 'A safety bulletin on the hazards of nitrogen asphyxiation describes an incident in which a cylinder of pure nitrogen was mistakenly delivered to a nursing home along with the correct shipment of pure oxygen cylinders. The nitrogen cylinder had a nitrogen label partially covering an oxygen label and was fitted with nitrogen-compatible couplings. A maintenance employee removed a fitting from an empty oxygen cylinder and used it as an adapter to connect the nitrogen cylinder to the oxygen system. Pure nitrogen was thus delivered to nursing home residents, resulting in four deaths and six injuries' [4].

Even if the correct parts are fitted properly problems can occur due to foreign objects. These can be tools, rags, and other items used during the maintenance activity, which can cause blockages. Sometimes items used to prevent foreign objects entering a system become the blockage themselves if not removed. Some industries apply procedures to establish 'clean conditions' that can be highly effective at preventing foreign objects entering a system. Using

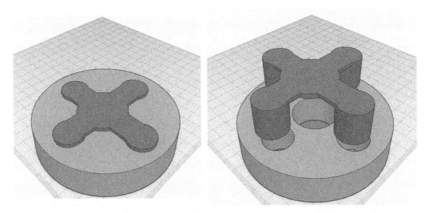

FIG. 4.15    Example of a flange end cap that cannot be left in accidentally.

appropriate methods of capping pipework and other flanges including those on valves that stop foreign objects entering the item but do not allow reassembly until they have been removed can be used [28]. The following image shows an example of flange cap that cannot be left in place because it fills the bolts holes in the flange (Fig. 4.15).

In some cases maintenance involves specialist skills to achieve the designed effects.

A vessel failed at the Marcus Oil and Chemical site in Houston, Texas (USA) in 2004 [29]. This caused structural damage and glass breakage occurred up to 400 m (quarter of a mile from the facility), injuring local residents. Also, three firefighters were slightly injured during the incident. The investigation concluded that a repair weld did not meet generally accepted industry standards for pressure vessel fabrication. Faults included failure to grind off the flame-cut surface that was welded. Also, the welds did not penetrate the full thickness of the vessel head and contained excessive porosity from gas bubbles. These defects significantly degraded the strength of the weld. Marcus Oil "did not use a qualified welder or proper welding procedure … and … did not hydrostatically pressure test the vessels after the welding was completed, a critical postwelding performance test" [29].

### 4.4.2.3 Repairing and overhauling equipment

Maintenance often involves repairing or overhauling equipment and other items. This is sometimes as simple as cleaning and replacing consumable items such as washers and seals. In other cases it can involve specialist reconditioning and testing, which may be performed on site but often involves sending the item away. Kletz highlighted that the practices followed in workshops when repairing and overhauling equipment require close scrutiny [7].

---

### Text Box 4.33

A fire started with a leak of hydrogen and olefins from the body of a high pressure valve of an unusual design, operating at a pressure at 250 bar (3600 psi). The leak occurred at a spigot and recess joint between the two halves of the valve body; the gap between them was too large and some of the joint material was blown through it. The diametral clearance should have been 0.05 mm (0.002 in.) or less but in fact varied from 0.53 to 0.74 mm (0.021 to 0.029 in.) and was too large to contain the 0.8 mm (1/32 in.) compressed asbestos fibre gasket.

The valve concerned, and others like it, had been overhauled and repaired by a specialist workshop which had undertaken such work for decades. There were no written procedures but the 'old hands' knew what was the standard required. However, most of these men, including the supervisors and inspectors, had retired and standards had slipped.

We should never ask a workshop to simply 'overhaul', 'repair' or 'recondition' equipment but should agree with them precisely what is to be done and the standard required. Quoting the original Drawing or Standard is not sufficient if they do not specify allowable wear down tolerances.

Relief or pressure safety valves (PSV) are a good example of equipment that is routinely sent away for overhaul and inspection. They are often our last line of defence in avoiding catastrophic damage to equipment and major accidents caused by high or low pressure and we need them to be reliable. PSVs are precision items and maintaining them is a complex activity and other issues arise due to associated activities including transport and storage. People often assume that sending their PSVs away to a self-identified specialist is enough but in practice there are many opportunities for things to go wrong [28].

Kletz used the Adventures of John series to illustrate how making sure the correct equipment is returned after a service can be ensured (Fig. 4.16) [8].

### 4.4.2.4 Replenishing fluids

Some routine maintenance or servicing involves topping up lubrication, hydraulic fluid, and seal oils, and applying grease. Also, this is often required after other maintenance has been carried out. Neglecting these steps will often have longer term consequences such as increasing wear of moving parts but using the wrong fluids can have more significant effects. For example, copper is a known sensitiser for ammonium nitrate so lubricating oils and

**FIG. 4.16**  Adventures of John—replacing the wrong relief valve can mean a vessel is not protected against over pressure [8].

greases that contain copper must never be used where contact is possible. Kletz gave the following example to highlight how apparently insignificant changes to the fluids made with good intent can have unexpected effects [4].

---

**Text Box 4.34**

Several years ago the UK Mines Inspectorate discouraged the use of mineral oils as hydraulic fluids underground. Officials were concerned that leaks would ignite, and a change was made to 'safer' less combustible fluids based on emulsions of phosphate esters. Unfortunately, these fluids are poor lubricants, and they caused excessive wear and overheating. This caused vaporisation of the water in the emulsions and further damage; the new fluids also attacked the rubber seals, and it was therefore agreed that, on balance, the conventional oils are safer.

---

People must know what fluids are approved for use and obtain formal authorisation if any change is being considered. Mixing different types or grade of fluid (i.e. topping up with a different fluid to the one that is currently in place) should be identified as a potentially significant change.

### 4.4.3 Monitoring maintenance work in progress

Maintenance that continues for any length of time should be monitored closely to ensure that it is going according to plan and to confirm that any changes to scope are being managed appropriately.

#### 4.4.3.1 *Daily checks and site inspections*

Checks of the maintenance worksite should be carried out frequently. Issues to look out for include:

- Cleanliness of the site including evidence of contamination and control of potential foreign objects;
- Conditions that spare parts, tools, and other materials are being kept in;
- Evidence that agreed RAMS and/or procedures are being adhered to;
- Evidence that permits to work are correct and up to date;
- Identifying any changes to the work party and whether these affect the competence to do the job safely and to the required standard.

There can be cultural issues with this type of monitoring. Maintenance personnel may not accept non-specialists commenting on their work and other teams may feel that it is not their responsibility. But errors made during maintenance can be very difficult to detect after the work is complete and can have significant consequences some time later.

#### 4.4.3.2 *Permit to work audits*

Formal audits should be carried out on a regular basis to ensure maintenance risk controls are effective in practice. The significant role of permit to work in controlling maintenance means that auditing its effectiveness gives very good insight into how the overarching system is functioning.

Guidance from the UK's Health and Safety Executive includes a checklist for the assessment of permit to work systems [26]. It encourages the following aspects to be examined critically:

- Policy—are objectives clear and is the system flexible enough to cover all types of activity that may occur?
- Control—are responsibilities for managing the system clearly defined?
- Communication—does the system support communication with everyone involved?
- Training and competence—does the system ensure that everyone who has a role in permit to work is competent in that role?
- Planning and implementation—does the system include clear rules about the preparations required for work and how it should be controlled including stopping work if hazards arise?
- Measuring performance—are routine checks and inspections of work taking place and are the findings being responded to.

The guide states that audits should be carried out by people not normally employed at the site wherever possible and that a procedure for the reviews is required [26].

### 4.4.3.3 *Responding to changes during maintenance*

Lots of assumptions have to be made when planning maintenance. It is common for the scope of the work to change part way through. These changes may introduce new hazards or make the planned precautions ineffective. The challenge is that the people carrying out the work are not always aware of the significance of changes because they occur gradually over time.

In the days preceding the Tosco Avon refinery fire that killed four people there had been a number of leaks that indicated that the isolations were not effective at keeping flammable materials in operating plant away from the pipework being replaced [9]. Tosco personnel failed to reassess the hazards of the work or to take measures to ensure that it could continue safely. In this case, the leaks should have been a trigger to say the plant needed to be shutdown and hazards removed. The company did have a policy that allowed workers to halt work that they thought was unsafe. However, these people would often be "be subject to a variety of external pressures to get the job done."

Managers and safety specialists should be involved in decision making and oversight of work. There should be effective safety reviews before work starts. Assuming work will proceed exactly as planned and then relying on people working at the sharp end to highlight issues is not good enough. We all have a duty to question what is difficult or different to the plan, in order to encourage those at the sharp end to openly discuss issues, rather than think they have to solve them alone, so we can actively help them resolve them in a safe way.

## 4.4.4 Returning to service

Restarting systems in order to return them to service is one of the most hazardous activities undertaken. This was graphically demonstrated by the BP Texas City explosion in 2005 that killed 15 people and injured 180 [30] (Section 3 of Appendix). There are inherent risks of introducing hazardous materials and energy to a system, which are further increased if errors have occurred during maintenance.

### 4.4.4.1 Checking completed work

Kletz used the Clapham Junction railway accident to illustrate how errors made by people carrying out maintenance can result in accidents when the system is returned to service [8].

---

**Text Box 4.35**

The Clapham Junction railway accident
The immediate causes of this 1989 accident, in which 35 people were killed and nearly 500 injured, were repeated non-compliances and a slip type human error. But the underlying cause was failure of managers to take, or even to see the need to take, the action they should have taken.

The non-compliances, errors in the way wiring was carried out by a signalling technician (not cutting disused wires back, not securing them out of the way and not using new insulation tape on the bare ends), were not isolated incidents; they had become his standard working practices. The official report stated 'That he could have continued year after year to continue these practices, without discovery, without correction and without training illustrates a deplorable level of monitoring and supervision within British Rail which amount to a total lack of such vital management actions'.

In addition, the technician made 'two further totally uncharacteristic mistakes' (disconnecting a wire at one end only and not insulating the other, bare end), perhaps because 'his concentration was broken by an interruption of some sort' and because of 'the blunting of the sharp edge of close attention which working every day of the week, without the refreshing factors of days off, produces'. 'Any worker will make mistakes during his working life. No matter how conscientious he is in preparing and carrying out this work, there will come a time when he will make a slip. It is those unusual and infrequent events that have to be guarded against by a system of independent checking of his work'. Such a system, a wire count by an independent person or even by the person who did the work, was lacking.

---

The accident report highlighted that the technician's supervisor was supposed to check the work. Kletz explained why this proved to be ineffective [8].

---

**Text Box 4.36**

The supervisor was so busy working himself that he neglected his duties as supervisor.
The original system of checking was a three-level one: the installer, the supervisor and a tester were supposed to carry out independent checks. In such a system people tend after a while to neglect checks as they assume that the others will find any faults. Asking for too much checking can increase the number of undetected faults.

---

Kletz used the following accident to show one of the reasons why checking can be ineffective [8].

---

**Text Box 4.37**

A man was asked to lock off the power supply to an electric motor and another man was asked to check that he had done so. The wrong breaker was shown on a sketch given to the men but the right one was shown on the permit and the tag and was described at a briefing. The wrong motor was locked off. The report on the incident states, "In violation of the site conduct-of-operations manual, which requires the acts of isolation and verification to be separated in time and space, the isolator and the verifier worked together to apply the lockout." Both isolator and the verifier used the drawing with the incorrectly circled breaker to identify the breaker to be locked out: neither referred to the lockout/tagout permit or the lockout tags, although both initialled them.

---

Although the numbers used by Kletz cannot be validated, his numerical approach does add some emphasis to the issue that checking may not reduce risks [8].

---

**Text Box 4.38**

If a man knows he is being checked, he works less reliably. If the error of a single operator is 1 in 100, the error rate of an operator plus a checker is certainly greater than 1 in 10,000 and may even be greater than 1 in 100 - that is, the addition of the checker may actually increase the overall error rate. A second man in the cab of a railway engine is often justified on the grounds that he will spot the driver's errors. In practice, the junior man is usually reluctant to question the actions of the senior.

---

Checking can provide some benefit if the checker works completely independently of the people who are doing the work being checked. Kletz claimed that this had been calculated [8].

---

**Text Box 4.39**

If two men can swop jobs and repeat an operation then error rates come down. For example the calibration of an instrument in which one man writes down the figures on a check-list while the other man calls them out. The two men then change over and repeat the calibration. The probability of error was put at $10^{-5}$.

Requiring a man to sign a statement that he has completed a task produces very little increase in reliability as it soon becomes a perfunctory activity.

---

Achieving independent checking is very difficult or impossible in practice. Also, it must be recognised that lots of maintenance activities cannot be checked after completion because they are not visible or are hidden inside equipment. It is quite common for companies to include checking steps in their procedures, often as a result of errors that have happened in the past, but in many cases the checks are signed off as complete without any active checking taking place.

### 4.4.4.2 *Signing off paperwork*

Paperwork is used to control maintenance work and signing it to show completion of work is usually required. However, accidents have happened because people have not realised that completion of work, which may be indicated on paperwork does not necessarily mean it is safe to proceed with other actions required to return plant to service.

Kletz told the following story where the full scope of maintenance work was not completed before plant was de-isolated [13].

---

**Text Box 4.40**

A similar incident occurred on a solids drier. Before maintenance started, the end cover was removed, and the inlet line was disconnected. When maintenance was complete, the end cover was replaced, and at the same time the inlet pipe was reconnected. The final job was to cut off the guide pins on the cover with a cutting disc. The atmosphere outside (but not inside) the drier was tested, and no flammable gas was detected. While cutting was in progress, an explosion occurred in the drier. Some solvent had leaked into the inlet pipe and then drained into the drier. The inlet line should not have been reconnected before the guide pins were cut off.

---

Risks are increased if multiple maintenance tasks are carried out and/or they are covered by multiple pieces of paperwork. Cross-referencing of permits is important at issue but cannot be relied upon, especially if permits have been issued at different times. Isolation certificates are often signed off when locks are removed but this does not mean that the valve or other isolation point has been physically returned to its normal status.

Removal of an isolation was identified as one of the causes of the Piper Alpha disaster (- Section 8 of Appendix). The operators did this to start a pump without realising its relief valve had been removed for maintenance. The underlying cause was that the pump isolation was not cross referenced with the removal of the relief valve. This was compounded by poor communication at shift handover and failures in the permit to work system, particularly in the management of suspended work [31].

### 4.4.4.3 *Tidying up*

There are lots of small tasks to perform after maintenance is complete that are often seen as relatively trivial and so may not receive the attention they deserve. They include general housekeeping, which if overlooked can result in accumulation of debris on site, which can be a hazard to people and even a fire risk.

Replacing insulation lagging that was removed to allow maintenance is a common omission and can have serious effects on the process. Also, failure to reinstate labels that may have been removed or obscured during maintenance will contribute to future problems with identifying equipment that can contribute to significant incidents as described earlier.

Failure to take responsibility for these trivial tasks is often the problem. Maintenance personnel feel that they are under pressure to finish work and move on. Operations personnel feel that these tasks should have been part of completing the maintenance. The general view is that 'someone else' will come along and deal with it at some time, but no one really knows who that someone is.

#### 4.4.4.4 Inhibits and overrides

A particularly critical step when returning to service is to make sure all safety devices that may have been inhibited, overridden, or disabled in some way are reinstated. Keeping a detailed log of every inhibit, override, and other actions that may disable a safety device when they happen is vital but should not be relied on because recording errors occur frequently. Line-walks and system checks should be carried out before every start-up. Some care is required to distinguish maintenance inhibits and overrides from inhibits and overrides used as part of start-up procedures.

Kletz used a fatal boiler accident to highlight this issue [8].

---

### Text Box 4.41

An official report described a boiler explosion which killed two men. The boiler exploded because the water level was lost. The boiler was fitted with two sight glasses, two low level alarms, a low level trip which should have switched on the water feed pump and another low level trip, set at a lower level which should have isolated the fuel supply. All this protective equipment had been isolated by closing two valves. The report recommended that it should not be possible to isolate all the protective equipment so easily.

We do not know why the valves were closed. Perhaps they had been closed for maintenance and someone forgot to open them. Perhaps they were closed in error. Perhaps the operator was not properly trained. Perhaps he deliberately isolated them to make operation easier, or because he suspected the protective equipment might be out of order. It does not matter. It should not be possible to isolate safety equipment so easily. It is necessary to isolate safety equipment from time to time but each piece of equipment should have its own isolation valves, so that only the minimum number need be isolated. Trips and alarms should be isolated only after authorization in writing by a competent person and this isolation should be signalled in a clear way - for example, by a light on the panel, so that everyone knows that it is isolated.

In addition, although not a design matter, regular checks and audits of protective systems should be carried out to make sure that they are not isolated. Such surveys, in companies where they are not a regular feature, can bring appalling evidence to light. For example, one audit of 14 photo-electric guards showed that all had been bypassed by modifications to the electronics, modifications which could have been made only by an expert.

---

#### 4.4.4.5 Leak and pressure testing

Leaks are common during start-up. This can be due to errors made during maintenance but can also be the result of thermal changes or seals relaxing whilst depressurised. Leak testing with inert fluids should always be considered before introducing any hazardous materials.

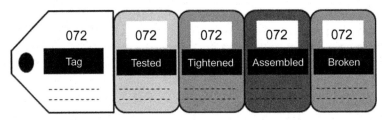

**FIG. 4.17**  Example of a multi-part tag for tracking the status of broken joints.

Joints broken as part of the maintenance scope should receive particular attention during leak testing. Logging every joint that is broken is important so that people know which to check when returning to service. But again errors are easy to make, especially given issues with labelling that have been discussed previously.

Some companies use multi-part tags that are attached when a joint is broken. Parts are removed when the joint is remade, when it has been torqued, leak tested, and finally tested on return to service. Some companies leave a part on the joint so that future breaks can be cross referenced (Fig. 4.17).

Making joints can appear to be a simple task and sometimes is overlooked as a critical task. There are many things that can go wrong including fitting the wrong gaskets, fitting them the wrong way round or misaligned, using the wrong bolts, torqueing incorrectly, etc. People need to be trained and proven competent in all aspects of the specific types of joint they may encounter. Unfortunately there can be an attitude that it is not so important because if a leak is found the joint can just be tightened up, but this may not address the underlying errors made when making it.

The term pressure testing is sometimes used interchangeably with leak testing. It is actually a formalised method that has its own hazards because systems are taken to high pressures, typically up to 50% above the design pressure. Hydrotesting with a non-hazardous liquid is usually preferred but can require extra work to remove the liquid afterwards. Pneumatic testing with air or inert gas avoids this problem but is more hazardous because pressurised gases store far more energy than pressurised liquids. Whatever method is used, pressure testing should only be performed when a specific need has been identified and should be documented in a detailed safe system of work or procedure [32].

Seven people were killed in 2010 at a refinery in Washington (USA) when hydrogen gas mixed with naphtha was released from a heat exchanger during start-up after they had been cleaned [5]. The technical cause was High Temperature Hydrogen Attack (HTHA). However, the accident itself and the number of casualties were influenced by poor practices. It had become common practice on the plant to use steam lances to control leaks during start-up. Continuing with known leaks by itself is a deviation from normal start-up and the plant should have seen this as a significant change in risk level. Also, it required a significant number of people to be present in an area that clearly proved to be hazardous.

### 4.4.4.6 Pre-start-up safety review (PSSR)

It is easy for operations teams to feel pressurised to restart plant after it has been shutdown for maintenance within a certain time or to achieve a specific date and time deadline. This can lead them to commence the restart before all the equipment and systems are available, or to overlook some pre-start checks. Also, they may plough on with the start-up despite encountering technical issues. Whilst having good start-up procedures and clear rules about compliance can be effective at minimising this risk, a formal Pre-Start-up Safety Review (PSSR) can reduce the risk further by passing responsibility for deciding whether to proceed with the start-up to management.

Two people were killed and eight injured at the Bayer CropScience facility in Institute, West Virginia (USA) in 2008, when a runaway reaction occurred during start-up of the methomyl unit after a maintenance outage [33]. During the investigation operations personnel acknowledged that there was not a specific deadline to complete the start-up but they were aware of dwindling stockpiles of materials. They commenced the start-up before the full scope of the planned outage maintenance had been completed including replacement of a damaged valve, faulty heat tracing, and modifications to process control. Also they proceeded with the start-up even after encountering electrical problems with a centrifuge, which meant that only one was operational even though the procedure stated that two were required, operating in parallel. Operators did not review laboratory test results, skipped a critical start-up safety prerequisite to fill and heat-up a solvent system, and bypassed a minimum temperature interlock. The result was the runaway reaction that caused the accident. The investigation found that a PSSR checklist had been completed but it had a number of critical omissions. Also, items had been indicated as being completed when they clearly had not been.

The purpose of the PSSR is to confirm that the start-up will be safe and straightforward. Items that can be overlooked, which should be focussed on during a PSSR, include making sure that:

- All valves, instruments, and equipment are accessible and any scaffolding or other items used during maintenance have been removed;
- All positive isolations (blank flanges, removed spools) should have been reversed, because it can be very complex and hazardous to do this if left too late;
- All required utilities and other services are available;
- Safety critical valves that are normally locked open or closed are secured in their operating condition;
- Temporary workstations set up for maintenance should have been removed or at least all work at them should have been stopped; and
- If changes have been made during maintenance the requirements of management of change should be confirmed as complete.

### 4.4.4.7 *Start-up*

In many ways the requirements at start-up are similar to shutdown. Detailed procedures should be followed that cover the specific start-up scenario. An instruction to simply follow the shutdown procedure in reverse is definitely not good enough.

Key objectives during start-up are to remove air, water, and any other fluids that may have entered during maintenance that are incompatible with the process. Pressurising with process fluids should be carried out in stages so that any leaks can be detected early, meaning the hazard is reduced. Equipment should be heated up gently to minimise thermal stresses. Organisationally there needs to be a sufficient number of people with the required competencies to perform the start-up including supervision.

Kletz highlighted why return to service and start-up have to be conducted in a very controlled way, and that everyone involved needs to understand the process [8].

---

### Text Box 4.42

An explosion occurred in a new storage tank still under construction. The roof was blown off and landed, by great good fortune, on one of the few pieces of empty ground in the area. No-one was hurt. Without permission from the operating team, and without their knowledge, the construction team had connected up a nitrogen line to the tank. They would not, they said, have connected up a product line but they thought it would be quite safe to connect up the nitrogen line. Although the contractors closed the valve in the nitrogen line it was leaking and a mixture of nitrogen and flammable vapour entered the new storage tank. The vapour mixed with the air in the tank and was ignited by a welder who was completing the inlet piping to the tank.

The contractors had failed to understand that:

- The vapour space of the new tank was designed to be in balance with that of the existing tank, so the nitrogen will always be contaminated with vapour.
- Nitrogen is a process material, it can cause asphyxiation, and it should be treated with as much care and respect as any other process material.
- No connection should be made to existing equipment without a permit-to-work, additional to any permit issued to construct new equipment. Once new equipment is connected to existing plant it becomes part of it and should be subject to the full permit-to-work procedure.

---

Removal of air is a critical step but if carried out with process fluid it is very important that it is vented to a safe place. US Chemical Safety and Hazard Investigation Board published a safety bulletin in 2009 [34] about the dangers of purging gas piping into buildings. It cited five serious accidents where purge gas had caused explosions that caused serious damage and injured nearly 100 people in total, including one where three workers were killed. Unfortunately, 5 months after this bulletin was published six people were killed at the Kleen Energy power plant in Middletown, Connecticut (USA) caused by gas purging [35]. The consequences of this incident were so much greater because people had not been excluded from the area whilst this hazardous activity had been taking place.

Kletz used the Adventures of Joe series to illustrate situations where removal of water is critical, particularly when introducing hot fluids including steam (Fig. 4.18) [8].

Joe put steam into a cold main without opening the steam trap by-pass

**FIG. 4.18** Adventures of Joe—introducing hot fluids is a hazardous activity [8].

## 4.5 Conclusions

Kletz shared his experience of looking at how maintenance is managed in practice [1].

---

**Text Box 4.43**

I visited plants, looked at permits-to-work and at actual isolations, and found that on the whole the new rules were followed, though sometimes people were not clear precisely what they should be doing. However, I found other things wrong. Although isolation was, on the whole, satisfactory, identification was poor. Maintenance workers sometimes opened up the wrong equipment. I had a long struggle to persuade Olefin Works in particular that equipment under repair, if not permanently labelled, should be identified by a numbered tag. Oil Works had started tagging a few years earlier. To fit slip-plates plants have to be designed so that there is room for the slip-plate and this meant convincing the design department that they should design accordingly. This was not difficult as my Oil Works colleague Joe Heaton was now in charge of piping design. Other Divisions knew little of the North Tees fire and had not changed their standards.

---

Unfortunately maintenance rarely receives the attention it requires. When budgets are being cut it is usually the top of the list. Some managers have received rapid promotion as the result of slashing maintenance costs. If they work for a large multi-national they do not usually stay at the site long enough to still be there when the problems created by reduced maintenance, which can take a couple of years to develop, start to bite.

Often problems occur during maintenance because it has not been properly considered during design. Whilst we have had some success at having operations representatives in project teams we very rarely have anyone with responsibility for maintenance.

In 2014 an incident occurred on a Boeing 747 aircraft. Following take-off from Gatwick airport the crew received warning of hydraulic leaks [36]. They prepared to return to the airport but were prevented from fully deploying the landing gear. The investigation found that an actuator for opening the landing gear door had been installed incorrectly during maintenance carried out immediately before the flight. This meant the door did not open properly and interfered with the landing gear. The investigation found that no physical arrangement was in place to prevent the actuator being installed incorrectly by 180 degrees and it was not easy to confirm it had been installed correctly. Also, the task required a special tool that was often not available because the task was performed very infrequently. In this case there were no causalities but the incident could have been prevented if maintenance requirements had been considered properly during design.

People carrying out maintenance have responsibility and should be trained to recognise the hazards they face in the course of their work. This will help improve their personal safety, situational awareness, and may help prevent major accidents [6]. However, managers and leaders have greater responsibility and should have proven competence in assessing the risks of maintenance work so that they can anticipate the hazards that are inherent in the system or that could be created during the work [6].

# References

[1] T. Kletz, By Accident—A Life Preventing Them in Industry, PFW Publications, 2000.
[2] US Chemical Safety and Hazard Investigation Board, Key Lessons From Preventing Incidents When Preparing Process Equipment for Maintenance, CSB Safety Bulletin, 2017.
[3] BP Process Safety Series, Safe Ups and Downs for Process Units, Institute of Chemical Engineers, 2006.
[4] T. Kletz, Plant Design for Safety—A User Friendly Approach, Hemisphere Publishing Corporation, 1991.
[5] A. Musthafa, Managing Risk in Major Maintenance—A Case Study on Fire and Explosions in the Process Industry, Loss Prevention Bulletin 268, 2019.
[6] P. Okoh, Maintenance-Related Major Accidents in the Metal Industry—The Combustible Dust Challenge, Loss Prevention Bulletin 268, 2019.
[7] T. Kletz, Lesson from Disaster. How Organisations Have no Memory and Accidents Recur, Institute of Chemical Engineers, 1993.
[8] T. Kletz, An Engineer's View of Human Error, third ed., Institute of Chemical Engineers, 2001.
[9] US Chemical Safety and Hazard Investigation Board, Tosco Avon Refinery Fire, CSB Investigation Digest, 2001.
[10] W.B. Rouse, R.M. Hunt, Human Problem Solving in Fault Diagnosis Tasks, US Army Research Institute for the Behavioural and Social Sciences, 1986.
[11] K.P. Block, Looking back at the Phillips 66 explosion in Pasadena, Texas: 30 years later, Hydrocarb. Process. (2019). https://www.hydrocarbonprocessing.com/magazine/2019/october-2019/special-focus-plant-safety-and-environment/looking-back-at-the-phillips-66-explosion-in-pasadena-texas-30-years-later. (Accessed 23 March 2020).
[12] Health and Safety Executive, HSG 253—The Safe Isolation of Plant and Equipment, second ed., (2006).
[13] T. Kletz, What Went Wrong?—Case Histories of Process Plant Disasters and How They Could Have Been Avoided, fifth ed., Elsevier, 2009.
[14] A. Brazier, Degrees of Separation, The Chemical Engineer, 2019.
[15] A. Brazier, Double Block and Bleed—It's More Complicated Than You Think, Loss Prevention Bulletin, 2016.
[16] K. van Gelder, Learning from reactive chemicals incidents at dow, Chem. Eng. Trans. 75 (2019).

[17] Health and Safety Executive, Chevron Pembroke Amine Regeneration Unit Explosion, 2 June 2011—An Overview of the Incident and Underlying Causes, (2020).

[18] European Union Network for the Implementation and Enforcement of Environmental Law, Explosion of a Sulphuric Acid Tank. ARIA Report 36628, (2011).

[19] US Chemical Safety and Hazard Investigation Board, Factual Investigative Update. Husky Superior Refinery Explosion and Fire, April 26, 2018 CSB, 2018.

[20] International Association of Oil & Gas Producers, Safety Performance Indicators—2013 Data. Fatal Incidents Report, (2014).

[21] A. Brazier, Interlocking Isolation Valves—Less is More, Institute of Chemical Engineers Hazards 27, 2017.

[22] T. Kletz, Critical Aspects of Safety and Loss Prevention, Butterworths, 1990.

[23] US Chemical Safety and Hazard Investigation Board, Thermal Decomposition Incident. CSB Investigation Report, BP Amoco Polymers, Inc., 2002

[24] BBC News, Margam Rail Workers' Deaths: 'No Formal Lookout' Appointed, https://www.bbc.co.uk/news/uk-wales-50013208, 2019.

[25] Health and Safety Executive, Train Derailment at Potter Bar 10 May 2002. A Progress Report by the HSE Investigation Board, (2003).

[26] Health and Safety Executive, Guidance on Permit-To-Work Systems, HSG 250, 2010.

[27] Health and Safety Executive, Common Topic 3 Safety Critical Communications, Inspectors Toolkit, 2005.

[28] A. Brazier, Maintaining Bursting Discs and Pressure Safety Valves—It's More Complicated Than You Think, Loss Prevention Bulletin, 2019.

[29] US Chemical Safety and Hazard Investigation Board, Polyethylene Wax Processing Facility Explosion and Fire. CSB Case Study, (2006).

[30] US Chemical Safety and Hazard Investigation Board, Refinery Explosion and Fire, BP Texas City, CSB Investigation Report, 2007.

[31] A. Brazier, Shared Isolations, Loss Prevention Bulletin, 2018.

[32] Health and Safety Executive, Safety Requirements for Pressure Testing. HSE Guidance Note GS4, fourth ed., (2012).

[33] US Chemical Safety and Hazard Investigation Board, Pesticide Chemical Runaway Reaction Pressure Vessel Explosion, CSE Investigate Report, 2011.

[34] US Chemical Safety and Hazard Investigation Board, Dangers of Purging Gas Piping Into Buildings, CSB Safety Bulletin, 2009.

[35] US Chemical Safety and Hazard Investigation Board, Natural Gas Explosion Kleen Energy, February 7, 2010CSB Public Meeting, 2010.

[36] Air Accident Investigation Branch, Bulletin 10/2015, (2015).

# Control of modifications

## 5.1 Introduction

Kletz identified in his autobiography that modifications and other changes are a source of risk [1].

---

**Text Box 5.1**

Many accidents have occurred because changes were made in plants or processes and these changes had unforeseen side effects.

---

Kletz shared the following story [2].

---

**Text Box 5.2**

"Brunel's ship, Great Britain, completed in 1846, was the first iron, screw-propelled, ocean-going steamship ever built and at the time the largest ship ever built. On its maiden voyage from Liverpool to the Isle of Man it missed the island and ran aground on the coast of Northern Ireland. No-one foresaw that the iron in the ship would affect the compass. (No accident has a single cause and there was also an error in the charts). Once the hazard was recognised it was easy to avoid by mounting the compass on the mast and viewing it through a periscope."

---

All types of change can have safety implications. The potential for making errors when planning changes is covered separately in the chapter on human error. This chapter is focussed on issues with failing to recognise or plan for the effects a change will have on the system including the people who operate and maintain it.

## 5.2 Different types of change

### 5.2.1 Minor modifications

Kletz provided the following definition of 'Minor Modifications' [3]:

---

**Text Box 5.3**

This is a term used to describe modifications so inexpensive that either they do not require formal financial sanction or the sanction is easily obtained. They therefore may not receive the same detailed consideration as a more expensive modification.

---

People often fail to recognise that what they are doing is actually a modification or feel that the scale of change is so small that the risks must be negligible. Any modification can have unintended effects and these can be difficult to predict, especially for different modes of operation. Also, individual minor modifications may not introduce any significant risk but over time a number of them can combine to fundamentally change the way a system works. This is covered in more detail later (Gradual changes or 'creep').

### 5.2.2 Adding extra valves

Kletz gave the following example of a minor modification [3].

---

**Text Box 5.4**

A reactor was fitted with a bypass (Fig. 5.1A). The remotely operated valves A, B, and C were interlocked so that C had to be open before A or B could be closed. It was found that the valves leaked, so hand-operated isolation valves (a, b, and c) were installed in series with them (Fig. 5.1B). After closing A and B, the operators were instructed to go outside and close the corresponding hand valves. This destroyed the interlocking. One day, an operator could not get A and B to close. He had forgotten to open C. He decided that A and B were faulty and closed a and b. Flow stopped. The tubes in the furnace were overheated. One of them burst, and the lives of the rest were shortened.

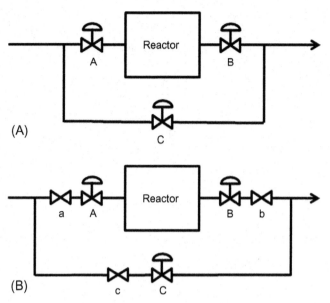

FIG. 5.1 (A) Original reactor bypass. (B) Modified reactor bypass.

This case illustrates how as systems get more complex it becomes more difficult to predict the impact of modifications. Devices like interlocks are often included in systems because designers feel that they remove the potential for human error. The reality is that any interlock can fail and they can usually be overridden in some way. Also, if the system's operation changes in any way or is used in an unusual mode of operation the interlock can require people to adapt and make changes to systems that might not have been necessary if the interlock had not been installed.

### 5.2.3 Other examples of minor modifications causing accidents

Kletz identified a number of minor modifications that contributed to cause an accident and provided the following list of examples [3].

## Text Box 5.5

1. Removing a restriction plate that limits the flow into a vessel and that has been taken into account when sizing the vessel's relief valve. A length of narrow bore pipe is safer than a restriction plate, as it is less easily removed.
2. Fitting a larger trim into a control valve when the size of the trim limits the flow into a vessel and has been taken into account when sizing the vessel's relief valve.
3. Fitting a substandard drain valve.

4. Replacing a metal duct or pipe by a hose.
5. Solid was scraped off a flaker—a rotating steel drum—by a steel knife. After the knife was replaced by a plastic one, an explosion occurred, probably because more dust was produced.
6. Without consulting the manufacturer, the owner of a set of hot tapping equipment made a modification: he installed a larger vent valve to speed up its use. As a result, the equipment could no longer withstand the pressure and was violently ejected from a pipeline operating at a gauge pressure of 40 bar (600 psi).
7. Making a small change in the size of a valve spindle and thus changing its natural frequency of vibration.
8. Changing the level in a vessel.

People need to have a thorough understanding of the system and its basis of design, and often a good imagination, to recognise all the possible effects of a modification.

### 5.2.4 Changing from a hose to hard piping to improve safety contributes to a leak of hydrofluoric acid

In 1998 approximately 10–20 kg of hydrofluoric acid (HF) leaked during tanker unloading [4]. HF is an extremely toxic chemical and so even small leaks like this are considered to be very serious. The immediate cause was a pinched gasket due to poor alignment of flanges at the connection between the plant and the ISO container tanker supplying the HF. Originally the delivery had used a flexible, steel braided hose. However, a safety audit suggested that the risks to personnel as a result of hose failures were too great and a 'flexible' hard piping system was introduced instead.

On the day of the leak the tanker had been left so that its connection point was slightly further away than usual. Operators found that they could make the joint by using a bit more effort. But this is how the gasket was pinched and the leak occurred. Unfortunately no form of leak test was conducted before the transfer commenced. However, personnel in the area were all wearing suitable personal protective equipment and were not harmed by the leak.

Multiple causes were identified in the subsequent investigation but one of the lessons learnt was [4]: "Any small plant change (even with seemingly good intention) should be reviewed in detail via HAZOP (or other 'what if' type methodology)."

## 5.3 Modifications made during maintenance

Kletz highlighted that modifications often happen whilst maintenance work is taking place with inadequate controls [3]:

**Text Box 5.6**

Even when systems for controlling modifications have been set up, modifications often slip in unchecked during maintenance. (Someone decides, for what he or she thinks is a good reason, to make a slight change.)

Better planning can help avoid this by making sure that the correct spare parts and consumables are available before maintenance work starts. However, maintenance personnel have to be alert and not assume that because something fits and looks the same as the item they have removed that it is exactly the same.

## 5.3.1 Pipework moved from the top to the bottom

Kletz used the following example to illustrate how slight changes sometimes occur during maintenance that can have a significant impact on how a system works [3] (Fig. 5.2).

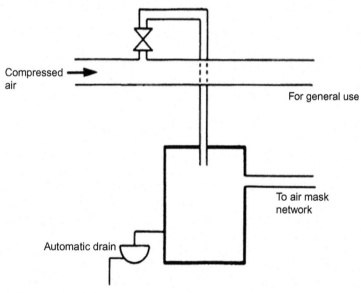

FIG. 5.2   Original arrangement of airlines.

**Text Box 5.7**

Many years ago, a special network of air lines was installed for use with air masks only. A special branch was taken off the top of the compressed air main as it entered the works.

For 30 years this system was used without any complaint. Then 1 day a man got a face full of water while wearing an air mask inside a vessel. Fortunately, he was able to signal to the standby man that something was wrong, and he was rescued before he suffered any harm.

Investigators found that the compressed air main had been renewed and that the branch to the breathing apparatus network had been moved to the bottom of the main. When a slug of water got into the main, it all went into the catchpot, which filled up more quickly than it could empty. Unfortunately, everyone had forgotten why the branch came off the top of the main, and nobody realised that this was important.

Kletz did not explain why the modification was made but it may just have been easier to access that location. Alternatively there may have been an issue with the site of the old branch and so it was easier to attach the new one to a clean area of pipe. People often think that pipework just has to connect one place to another and its exact arrangement does not matter. They are more likely to be fixated on creating a good joint, so that it doesn't leak. In this case the person attaching the new branch should have asked for permission to relocate the joint but equally the person overseeing the work would need to recognise that this was a modification that required further analysis. If the company had been focussed on keeping detailed and up to date records of 'as built' design people may have realised that as a minimum they would have had to ask for drawings to be updated after this work, which may have alerted someone to the potential problem, but unfortunately 'as built' drawings and other documents are often found to be hopelessly out of date or were never accurate in the first place.

### 5.3.2 The wrong spare parts

One of the challenges for maintenance personnel can be sources of the correct spare parts and identifying if the ones available are exactly the same or different in any way. Companies often state that 'like for like' replacements do not need to be evaluated but this term is not necessarily as clear as it may appear.

Over the life of a plant the suppliers of parts can change. Some go out of business and so a new part has to be found. People will often assume that, if it has the same specifications and ratings as the original, this is not really a modification—it is like for like.

Sometimes the supplier will be bought by another company and the same part is available but with a different description, part number, or packaging.

In other cases the original supplier may start to source its parts from a third party. The details may appear to be the same but the contents in the packet may have changed.

When the exact part is not available it can appear that a better part can be used. Sometimes this will be supplied at no extra cost as a sign of good will and sometimes the customer may not even be informed. Logic often suggests that having a higher rating can only be a positive change but this is not necessarily the case. For example:

- A more powerful motor may result in higher forces being applied to components causing them to fail;
- The original item may be intended to be a weak spot and using a stronger version allows more energy to build up in a system so that when a failure does occur it causes more damage;
- Using better materials (e.g. stainless instead of carbon steel) may result in dissimilar metals coming into contact causing increased rates of corrosion;
- Using better insulation may save energy but result in equipment being overheated.

Education of everyone involved in the supply chain right from purchaser to maintenance technician is an important part of making sure these issues do not occur. But the number of ways that problems can occur highlights why vigilance is required at all times.

Risks and control of modifications need to be highlighted in procurement and stores procedures and permits to work.

### 5.3.3 Example of a modification during maintenance

An Instrument and Electrical (I&E) technician received a work order to replace a faulty solenoid operating a block valve [5]. There were no solenoids of the correct size available in the stores but there was one of the next size up, which fitted and appeared to work in the same way. Unfortunately the valve in question was part of an emergency de-pressuring system (EDP) on a pressure vessel. The larger solenoid caused the vessel's outlet valve to close more quickly, meaning it reached its closed position sooner than the vessel's inlet valve. This meant that the vessel was still able to pressurise in a scenario where the EDP had been triggered to achieve exactly the opposite effect.

This incident highlights that a small change in physical and monetary terms can have a significant impact on a safety system. Avoiding this would have required the technician or the storeman to recognise that it was a change. Such decisions about small changes can have large consequential hazards. They are made at the 'coal face' on a day-to-day basis [5].

### 5.3.4 Continuing operations during maintenance

In 2011 a tank at a French refinery overflowed and $1000\,m^3$ of diesel was released to the bund [6]. One of the causes was a mistake during a previous maintenance activity when a valve identification tag had been fitted incorrectly. This meant that the operator opened a valve to a tank that was already full.

A high level alarm in the tank was relayed to the control room but the operators did not respond. The level instrumentation on the tank was being replaced but the new system was not yet fully operational. This meant that the control room operator was receiving many false alarms and failed to recognise that a genuine issue had occurred. If operations are to continue whilst modifications are being implemented it is essential that the potential effects are fully considered.

## 5.3.5 Temporary modifications

People often think that it is not worth the effort of obtaining authorisation for modifications that are only going to be in place for a short while, because they think that there will not be enough time for an accident to occur and it will take longer to make the assessment than the duration of the change. Unfortunately, this is not always the case.

### 5.3.5.1 *Flixborough accident*

Kletz used the Flixborough accident (see Section 6 of Appendix) as an example of the risks of temporary modifications [3]. In fact this accident is often quoted as the wake-up call for the process industry about why all modifications need to be managed.

---

### Text Box 5.8

The most famous of all temporary modifications is the temporary pipe installed in the Nypro Factory at Flixborough, in the United Kingdom, 1974. It failed 2 months later, causing the release of about 50 tons of hot cyclohexane. The cyclohexane mixed with the air and exploded, killing 28 people and destroying the plant.

At the Flixborough plant, there were six reactors in series. Each reactor was slightly lower than the one before so that the liquid in them flowed by gravity from No. 1 down to No. 6 through short 28 in. diameter connecting pipes (Fig. 5.3). To allow for expansion, each 28-in. pipe contained a bellows (expansion joint).

One of the reactors developed a crack and had to be removed. (The crack was the result of a process modification.) It was replaced by a temporary 20-in. pipe, which had two bends in it, to allow for the difference in height. The existing bellows were left in position at both ends of the temporary pipe.

The design of the pipe and support left much to be desired. The pipe was not properly supported; it merely rested on scaffolding. Because there was a bellows at each end, it was free to rotate or 'squirm' and it did so when the pressure rose above the normal level. This caused the bellows to fail.

No professionally qualified engineer was in the plant at the time the temporary pipe was built The men who designed and built it (design is hardly the word because the only drawing was a

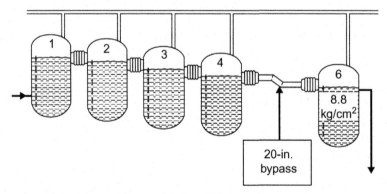

FIG. 5.3   Arrangements of reactors and temporary pipe at Flixborough.

full-scale sketch in chalk on the workshop floor) did not know how to design large pipes required to operate at high temperatures (150°C [300°F]) and gauge pressures (150 psi or 10 bar). Few engineers have the specialised knowledge to design highly stressed piping. But in addition, the engineers at Flixborough did not know that design by experts was necessary.

---

An expression that is commonly used to explain the issues that led to Flixborough is 'you don't know what you don't know'. It is very difficult for people to recognise the gaps in their knowledge and easy for them to assume that they know enough to get the job done. In everyday life we learn from our mistakes but we cannot allow this to happen when people are making decisions about hazardous systems.

Everyone has gaps in their knowledge, no matter how experienced or well qualified they are. But a key part of educating people to take responsible roles (e.g. Lead Engineer) is making sure that they know their limitations and when to ask for specialist support. As with all human factors issues there is a balance to strike here because people will not become skilled in their job if they always hand every difficult decision to someone else.

### 5.3.5.2 Evangelos Florakis Naval Base explosion

One of the problems with temporary modifications is that they can be in place for far longer than ever imagined when they are first installed. It is often possible to manage an unusual situation for a short time because people are aware that safety controls may be lower than would normally be accepted and so are particularly vigilant. However, as time passes this unusual situation starts to become normal and vigilance starts to degrade.

On 11 July 2011 a major explosion occurred at the Evangelos Florakis Naval Base in Cyprus [7]. Thirteen people died, including the Commander of the Navy, the Commander of the base, and six civilian firefighters who had been tackling the blaze prior to the explosion. Sixty-two people were injured and some 150 properties in the nearby villages of Mari and Zygi were damaged.

The accident happened in a stack of 98 shipping containers being stored in an open area of the site. These had been removed from a ship in January 2009 (30 months earlier) bound for Iran because they contained explosives and weapons in violation of United Nations sanctions. Despite the known hazards, multiple political and military discussions had failed to address this temporary situation and the containers had remained where they were.

The immediate cause of the explosion was identified as self-ignition in one of the containers triggered by the summer temperatures. Although the site was a military base and had procedures for safe storage and handling of ordnance, this was an unusual situation and the standard procedures were considered to not apply.

One of the key lessons from the investigation was that "The management system must ensure that a risk assessment is carried out when changes are made or new materials are introduced to a site. Depending on the scale of the change and the associated hazard, this might be achieved through a Management of Change procedure, which may require a formal Process Hazard Analysis."

Another lesson was "Beware of temporary arrangements that last for longer. This is especially relevant where there may be seasonal issues. In this case the explosives may have been safe if stored during the winter months, but not in the summer." A clear demonstration of why something that can be handled for a short time can become a major risk if the duration is extended.

### 5.3.6 Overriding safety devices

Overrides, inhibits, and bypasses are used more frequently than people often realise. Whilst this should be viewed as a temporary modification it is often considered to be an operational necessity and the need for evaluation and control is not always recognised. This can be a training or cultural issue but often it is a result of poor design and/or maintenance.

### 5.3.7 Plant and equipment start-up

Plant and equipment manufacturers can be very precious about their products and want to do what they can to protect them against damage. One way they do this is by including automatic protection devices that will trip the machine if it is operating outside of its normal envelope. The problem is that these features can often make it impossible to start the machine.

The solution is for operators to defeat the protection device during start-up. This may not normally be a problem if they are re-enabled within a very short time, once the machine is running. Of course people will sometimes forget to re-enable one or more of the devices that they have defeated and problems can occur during start-up or very soon afterwards whilst the device is still defeated. Machines do not trip and are damaged as a result.

Whilst the immediate risks are a concern, the effect that this practice has on wider perceptions can be even greater. People get used to the idea that protective devices need to be defeated to start a machine. If they have problems starting a machine because it keeps tripping they are liable to think that this new problem can be overcome by defeating more devices. Very quickly the routine can result in many more protective devices being defeated than is necessary, increasing the risk of machine damage because the levels of protection are reduced.

The ideal solution is to design systems so that they do not require any device to be defeated manually when performing routine tasks including start-up. Where this is not possible the provision of start-up overrides or inhibits that will automatically reset themselves when certain parameters are achieved or after an allotted time reduces the vulnerability to errors. However, even this requires evaluation.

In one case a hazardous scenario could occur due to cold temperature and a trip was installed. However, the plant was usually cold prior to start-up and so the trip had to be overridden. It was then decided that the trip should reset itself automatically as soon as a normal operating temperature was achieved, so that the operators did not forget. Unfortunately the way the process worked meant that a slug of warm material would come through quite soon and the trip would reset. However, a cold slug would then come through and the plant would trip before it had a chance to complete its full warm-up sequence. As a result operators became fixated on being ready to re-apply the override after it had reset in anticipation of the cold slug rather than on monitoring the rest of the plant whilst it was starting-up.

## 5.3.8 Avoiding a trip

Operators can feel under pressure to keep plant running and it is important to recognise that trips resulting in emergency shutdown of plant carry significant risk. With this in mind they can get into the habit of applying overrides or inhibits when they receive an alarm in the belief that they can get the situation under control before it becomes hazardous, but if they leave the protective device enabled it may cause the plant to trip or shutdown.

A plant experienced a loss of containment when a vessel overflowed. The investigation found that the operator had overridden a safety instrumented system (SIS) that would have closed the inlet valve to the vessel and prevented it from being overfilled. Further investigation showed that the high level alarm had activated several times during the shift and also the level transmitter was reading less than 100% when the vessel had overflowed.

The investigation concluded that level control of this vessel was problematic and it had become standard practice amongst all the operators to override the SIS whenever the high level alarm had been activated. Previously the operators had got the level under control quickly and had re-enabled the SIS once the alarm had cleared. On this occasion the control problem was much more severe and the normal actions to rectify it had been unsuccessful. The operator was paying full attention to the level and felt that because the transmitter was showing it to be less than 100% he still had time to re-establish control. Unfortunately, the transmitter was incorrectly calibrated (actually the result of a change to the process) meaning the operator had no way of knowing that it was about to overflow.

This is a good example of where routinely defeating safety devices not only increases risks but it also effectively hides system problems. If the operators had allowed the plant to trip due to high levels in this vessel it seems very likely that managers would have been willing to invest time and effort to rectify the underlying problems. However, because the operators had become adept at keeping the plant running managers were not aware of the problems.

## 5.4 Process modifications

Whilst it can be difficult to identify and control physical changes to plant and equipment, process changes can be even more subtle but equally risky. Given that plants often cycle between different states with different process conditions it can be difficult to understand that some changes are dangerous.

## 5.4.1 Smaller particles

Kletz shared this example of where the only change was the size of charcoal particles in a filter [3].

## Text Box 5.9

A hydrogenation reactor developed a pressure drop. Various causes were considered—catalyst quality, size, distribution, activation, reactant quality, distribution, and degradation—before the true cause was found. The hydrogen came from another plant and was passed through a charcoal filter to remove traces of oil before it left the supplying plant. Changes of charcoal were infrequent, and the initial stock lasted several years. Reordering resulted in a finer charcoal being supplied and charged without the significance of the change being recognised. Over a long period, the new charcoal passed through its support into the line to the other plant. Small amounts of the charcoal partially clogged the 3/4 in. (10 mm) distribution holes in the catalyst retaining plate (Fig. 5.4). There was a big loss of production. The cause of the pressure drop was difficult to find because it was due to a change in another plant.

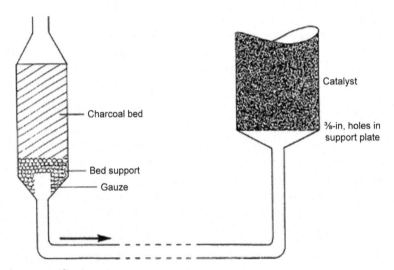

**FIG. 5.4**   Hydrogen purification system.

## 5.4.2 Good intentions

Kletz shared this story where switching off a fan to save energy had unforeseen consequences [3].

## Text Box 5.10

Three vacuum stills were fitted with steam ejectors and direct contact condensers. The water was cooled in a small cooling tower and recycled, and the small amount of vapour carried over from the stills dispersed in the tower. The tower was fitted with a fan, but to save electricity, the operators switched off the fan and found that they could get adequate cooling without it. However, the flammable vapours from the stills were no longer dispersed so effectively. This was discovered when an operator, about to light a furnace a few metres away, tested the atmosphere inside the furnace in the usual way—with a combustible gas detector—and found that gas was present. No one realised, when a site for the furnace was decided, that flammable vapours could come out of the cooling tower. Direct contact condensers are not common, but flammable vapours can appear in many cooling towers if there are leaks on water-cooled heat exchangers. After the incident, a combustible gas detector was mounted permanently between the furnace and the tower.

---

This is one of the challenges. We want people to use their initiative and we will generally congratulate them when it works out well. But everyone has to understand that any change can cause problems, possibly sometime after the change is made so it can be difficult to see the link between the change made and the problem that occurs as a result.

### 5.4.3 Change of raw material reactivity

In 2013 a chemical site in France experienced a runaway reaction during production of aluminium orthophosphate, which resulted in two tonnes of hot reagents being expelled from the reactor [6]. The investigation concluded that a change had occurred in the raw material. Although it was the correct chemical, the aluminium hydroxide being used at the time was more reactive than had previously been the case. This had been a planned change and some tests had been carried out, but they did not include any calorimetric measurements that would have identified the potential issue. The technicians had noticed an increase in reactivity but their observations had not been reported. Following this incident the company improved its approval process for new raw materials and the way it managed deviations.

### 5.4.4 Fundão and Santarém tailing dam disaster

On 5 November 2015 the Fundão dam at a Brazilian mine owned by Samarco Mineração SA ruptured releasing a large volume of toxic sludge into the Santarém river valley. Nineteen people were killed and the river was very badly polluted.

One of the main causes of the dam failure was that changes had occurred that resulted in less efficient water drainage [8]. Sand in the dam walls became saturated and started to behave more like a liquid than a solid. It was unable to withstand the weight exerted on it and failed.

## 5.5 Modifying written procedures

Written procedures play a critical role in the management of risks but simply writing how to perform a task does not mean the task will be performed as written. If it is decided that a written procedure should be modified it is important to determine how the changes will be communicated to the people who perform the task and how to make sure they actually modify the method they use.

### 5.5.1 Procedure change not accepted

Kletz described the following accident following a change to a procedure [2].

---

**Text Box 5.11**

An example of an accident caused by a change in procedures: an operator was asked to add several reactants to a pilot plant reactor. His experience told him that there would probably be a fume emission so he changed the order of addition. This caused a runaway reaction.

The operator should not have made the change without the authorisation of the chemist in charge. But equally, if the chemist had discussed the procedure with the operator, he would have learnt about the fume problem.

---

For all types of modification, not just procedures, it is important to understand how systems work in practice. Often the people operating or maintaining the system have learnt things that may have not been captured formally in any document. Failing to understand how systems really work is one of the main problems with managing the risks of modifications and change. The other issue highlighted here is that if people affected by a modification are not told what is happening and why, they may not realise or understand why they need to change the way they perform a task that they may have performed using the current method for many years.

### 5.5.2 Modifying procedures due to other modifications

A procedure may need to be modified if something else has changed such as equipment and plant, staffing levels, regulations, standards, or guidance. In these cases modifying the procedure and communicating the changes should be managed as part of the primary modification (e.g. plant, process, or organisation).

Unfortunately it is very common for procedural updates to be left until to the last minute or overlooked completely, often because there is pressure to return the plant to operation as quickly as possible. It gives the impression that procedures are not important enough to bother with.

### 5.5.3 Modifying procedures in their own right

A procedure may need to be changed because it is identified as being deficient in some way. For example it may be found that the procedure is inaccurate, out of date, or not practical to follow. Staff may feedback that it is confusing, ambiguous, or not in an appropriate format for its intended use. In these cases it is the modification of the procedure that should be the managed directly.

### 5.5.4 Document control

Document control is an essential part of any procedure management system. It ensures that only authorised changes are made and that only the most up to date version of a procedure is available. It needs to be applied to electronic master copies and any hard copies that may be retained for easy access.

Unfortunately document control processes are often too slow to react to changes and actually contribute to poor quality procedures. This can have a direct impact on people when they need a procedure to perform a task. It causes significant frustration and can lead to people thinking that feeding-back about procedure inaccuracies is not worthwhile. They may as well struggle on with what they already have because it is too much effort to get the procedures modified.

## 5.6 Software modifications

Software has become increasingly important to all aspects of work and life. As technology has advanced software has had to evolve. Also, as other changes occur (e.g. plant, process) there is often a need to update the software of operating and possibly safety systems. The sheer complexity and legacy issues mean that knowledge about how older software works is very sparse, making it very difficult to evaluate any planned modifications.

### 5.6.1 Leak of anhydrous ammonia

On 22 August 2010 a leak of anhydrous ammonia occurred at a frozen food store in Theodore, Alabama. Eight hundred people working in the vicinity were exposed, 32 were admitted to the local hospital suffering from the effects of serious exposure, and four of these were treated in the intensive care unit. The triggering event was a power failure at the site lasting for over 7 h. When power was restored a software 'glitch' actively switched a blast chiller from defrost cycle into refrigeration without first clearing the coils of the hot ammonia gas used to clear the ice. This caused hydraulic shock that ruptured pipework and allowed the anhydrous ammonia to escape. It was concluded that unauthorised modifications of the system software had been made sometime in the past but had been undetected during normal operation [9]. Recommendations included protecting systems with password access, using formal methods to review modifications such as Computer Hazard and Operability (CHAZOP), and thorough testing before any modification goes live using defined test protocols.

### 5.6.2 Boeing 737 Max

On 29 October 2018 a Boeing 737 MAX 8 operated by Indonesian airline Lion Air crashed into the Java Sea 12 min after take-off; 189 passengers and crew died. Less than 5 months later on 10 March 2019 the same type of aircraft, this time operated by Ethiopian Airways crashed near the town of Bishoftu 6 min after take-off; 157 passengers and crew died.

The two aircraft were both relatively new and of the same recent design. Early analysis quickly showed that the events leading up to the crashes were similar and a piece of flight control software introduced on the Boeing 737 MAX 8 was identified as a likely culprit [10].

An application known as the Manoeuvring Characteristics Augmentation System (MCAS) had been installed to assist the pilots to overcome difficulties caused by a tendency of the aircraft's nose to rise too much, which can result in a stall. It appears that this system was activated in the minutes leading up to both accidents when it was not actually required. It angled the aircraft's nose down when it was supposed to be climbing. Evidence suggests that the pilots responded to the problem in accordance with procedures and training but this did not work.

These accidents raised a number of questions but clearly the way the flight control software had been modified was one of the main concerns.

## 5.7  Changes in organisation

Whilst companies in the process industry have had some success at controlling modifications to plant and process, largely as a result of the lesson learnt from Flixborough, they have tended to struggle more with changes in organisation: when either the people or the way they are organised change in some way. These changes can mean that systems that previously worked well are not suited to the new organisation and can result in poorer control of hazards and higher likelihood of accidents. Typical reasons why management of organisational change can be ineffective include: the scope of assessment is too limited, transitional risks are overlooked, the change is dictated from the top level management, or it is assumed that the change is dictated by business needs (cannot be challenged) [11].

### 5.7.1  Incident at an ethylene plant

Kletz described an incident at an ethylene plant where cold liquid overflowed from a distillation column into a flare system, which failed due to low temperature embrittlement [3]. Fortunately the leak did not catch fire and no one was injured, but the potential consequences were very great. Kletz identified a number of short-term and long-term organisational changes that had contributed to this accident.

The incident happened as the plant was being started up after a planned maintenance turn-around. An error meant that the level instruments on the column had not been recommissioned. However, the operators were too slow to notice that they were not receiving any data about the level. When they finally realised there was a problem they failed to recognise that the column was already flooded (full of liquid).

One of the short-term changes identified by Kletz as a cause of the incident was that operators were working their normal 8 h shift pattern whilst the foremen and shift managers had adopted a 12 h shift pattern, with start and finish times that did not match the operators. In Kletz's words [3].

---

**Text Box 5.12**

This pattern of work destroyed the cohesion that had been built up over the years within each shift and lowered the competence of the team as a whole. A report in the local newspaper said that "A major influence over the behaviour of the operating teams was their tiredness and frustration." A trade union leader was quoted as saying that the management team members were more tired than the operators as they were working 12-h shifts.

---

However, Kletz was able to identify more fundamental changes that had occurred in the years leading up to this incident that had affected the way the plant was operated [3].

---

**Text Box 5.13**

About 7 years earlier, there had been a major recession in the industry. As in many other chemical companies, drastic reductions were made in the number of employees, at all levels, and many experienced people left the company or retired early. This had several interconnected results:

- Operating divisions were merged, and senior people from other parts of the company, with little experience of the technology, became responsible for the ultimate control of some production units.
- There was pressure to complete the turnaround and get back on line within 3 weeks. This pressure came partly from above but also from within the production and maintenance teams, as the members were keen to show what they could do. They should have aborted the shutdown[sic] to deal with the problems that had distracted everyone during the night but were reluctant to do so.
- There were fewer old hands who knew the importance, when there were problems, of having a look around and not just relying on the information available in the control room. A look around would have shown ice on the demethaniser column.
- Delayering had produced a large gap in seniority between the manager responsible for the ethylene plant and the person above him. This made it more difficult for the ethylene manager to resist the pressure to get back on line as soon as possible. Previously, an intermediate manager had acted as a buffer between the operating team and other departments, and he prevented commercial people and more senior managers from speaking directly to the startup team. In addition, he would probably have aborted the startup. Senior officers, not foot soldiers, order a retreat.

---

## 5.7.2 High profile accidents where change to the organisation contributed

It is probably fair to say that many major accidents occur because failures within the organisation lead to risk controls being degraded. Very often this will have been because of a change to the organisation. The following high profile and well known accidents provide us with some clear illustrations [12]:

- Hickson and Welch Ltd. (1992)—A change of company structure had resulted in the role of plant manager being eliminated. An area manager, with support from others, was supposed to have responsibility for the plant; however, they had become overloaded and had not been able to provide sufficient attention to plan the activities.
- Esso Longford (1998)—Engineers, who had previously been located on site and readily available to deal with queries and concerns and maintain a degree of oversight, had been moved off-site, resulting in communication becoming more formal and focussed, with specific questions being answered by engineers but less general discussion taking place.
- BP Grangemouth (2000)—Underlying causes identified as poorly planned and executed organisational and personnel changes. In particular, changes to responsibilities and competency required to deliver these had not been properly understood or managed.

- BP Texas City (2005)—The organisation had evolved over time to a position where responsibility for process safety resided with the refinery plant manager, but they had many other responsibilities including day-to-day performance of the plant. As a result, process safety had received low priority.
- Bayer Crop Science LLC (2008)—The company had reduced the number of technical advisors it employed without ensuring that it retained the necessary knowledge of hazards and risks. One result was that operational input to the capital project was provided by someone with no relevant operational experience of the plant, meaning that the design was not consistent with the existing plant.
- Deepwater Horizon/Macondo (2010)—The platform was being operated by a complex organisational structure resulting from changes in the way the offshore oil industry conducts its business. Contracting and subcontracting meant that people on-board worked for different companies and inevitably reported to different supervisors and managers, many of whom would have been located onshore or on other platforms. Also, commercial sensitivities probably meant that people did not share all the information they knew.

The number of incidents with organisational changes as an underlying cause should be a wake-up call to us all that we need to get better at managing the changes that happen almost continually in the industry.

### 5.7.3 Types of organisational change

Any organisational change can affect the way risks are managed. This can range from a change affecting a single person through to a major corporate restructuring at a global level. The following provides some examples [12] and potential impacts.

- Reducing staffing levels—unable to keep on top of routine workload and may not be able to cope with high demand situations including emergencies;
- Using contractors or outsourcing—reduced levels of control over how work gets done and possible loss of internal capability to manage issues (i.e. if the contractor is not available);
- Combining or splitting departments—confusion about roles and responsibilities and disruption to team work;
- Changing roles and responsibilities—potential gaps (no one is responsible) and demotivation of individuals (change to their job imposed on them);
- De-layering and adopting new team structures (e.g. self-managed teams)—loss of specialist skills and areas being overlooked (everyone is trying to cover everything).

Some of the potential issues with organisational change include: less resource being available to complete routine work and an inability to handle incidents, loss of corporate memory and difficulty to manage competence, poorer communication both up and down the management hierarchies and between different teams and departments working at the same level.

### 5.7.4 Managing organisational change

Even the most simple of organisational changes should be subject to some form of assessment, if only to determine that there is no evident risk. Therefore, all organisational changes should initiate the management of change process [13]. The types of questions to ask may include:

- Does the change impact a safety critical role?
- Does the change impact Emergency Response arrangements?
- Does the change impact responsibilities within the Safety, Health & Environmental procedures and instructions?
- Does the change impact any unique or specialist knowledge within the company?

When considering change, doing nothing can be more hazardous than the alternatives. Therefore, the risk assessment process should consider any risks associated with not making a change from the current situation [13]. The assessment should be a consultative process involving the people affected by the change (or their representatives), their line manager(s), and other support personnel who have a good working knowledge of the work processes affected by the change. It has been suggested that all organisational changes should be checked by a safety committee [11].

## 5.7.5 A system for managing changes to an organisation

Kletz provided some detailed guidance on how to control managerial change [3]. He was definitely ahead of his time on this subject as many companies still do not have effective systems for doing this.

### Text Box 5.14

As with changes to plants and processes, changes to the organisation should be subjected to control by a system that covers the following points

- Approval by competent people. Changes to plants and processes are normally authorised by professionally qualified staff. The level at which management changes are authorised should also be defined.
- A guide sheet or check list. Hazard and operability studies are widely used for examining proposed modifications to plants and processes before they are carried out. For minor modifications, several simpler systems are available. Few similar systems have been described for the examination of modifications to organisations.
- Each modification should be followed up to see if it has achieved the desired end and that there are no unforeseen problems or failures to maintain standards. Look out for near misses and failures of operators to respond before trips operate. Many people do not realise that the reliability of trips is fixed on the assumption that most deviations will be spotted by operators before trips operate. We would need more reliable trips if this were not the case.
- Employees at all levels must be convinced that the system is necessary or it will be ignored or carried out in a perfunctory manner. A good way of doing this is to describe or, better, discuss incidents that occurred because there was no systematic examination of changes.

    Some points a guide sheet should cover.

    Define what is meant by a change. Exclude minor reallocations of tasks between people but do not exclude outsourcing, major reorganisations following mergers or downsizing, or high-level changes such as the transfer of responsibility for safety from the operations or engineering director to the human resources director. Accidents may be triggered by people but are best prevented by better engineering.

Nearly half of the companies that replied to a questionnaire on the management of change said that they included organisational change under this rubric. However, they may not include the full range of such changes.

Some questions that should be asked include the following

- How will we assess the effectiveness of the change over both the short and the long term?
- What will happen if the proposed change does not have the expected effect?
- Will informal contacts be affected (as at Longford)?
- What extra training will be needed, and how will its effectiveness be assessed?
- Following the change, will the number, knowledge, and experience of people be sufficient to handle abnormal situations? Consider past incidents in this way.
- If multiskilling is involved, will people who undertake additional tasks know when experts should be consulted?

Except for minor changes, these questions should be discussed by a group, as in a hazard and operability study, rather than answered by an individual. *None* or *not a problem* should not be accepted as an answer unless backed up by reasons. Any proposal for control of changes in an organisation should be checked against a number of incidents to see if it could have prevented them.

## 5.8 Gradual changes

Kletz highlighted that change often happens gradually as a result of multiple modifications. This means that a system's configuration can change significantly over time to be fundamentally different to its original design. Unfortunately, people do not always recognise this because every modification had a small impact at the time it was implemented.

### 5.8.1 Adjusting a design

Kletz provided this example where lots of small modifications led over time to a hazard [3].

### Text Box 5.15

In ammonia plants, the furnace tubes end in pigtails—flexible pipes that allow expansion to take place. On one plant, over the years, many small changes were made to the pigtails' design. The net effect was to shorten the bending length and thus increase the stress. Ultimately, 54 tubes failed, producing a spectacular fire.

Kletz did not say whether these modifications had been subject to a suitable review and authorisation process but experience shows that it can be difficult to appreciate the effect of multiple small modifications.

In 1999 a significant release of vinyl chloride monomer (VCM) occurred at a site in Victoria, Australia whilst the plant was returning to service after a maintenance shutdown/turnaround [14]. The VCM travelled to atmosphere via the utility drain system, which was supposed to handle water being transferred from vessels. The investigation found that multiple modifications had been made to the system over a 20-year period including facilities to inject steam for sparging water before discharge and then an air blowing system. A couple of years before the incident extra plant had been added to double production capacity, which had process links to the existing plant via the utility systems. Further analysis showed that records of the evaluations carried out for each modification were not readily available and studies that had been carried out (e.g. HAZOP) may have only considered facilities being added and not the potential impact on the existing plant. A conclusion was that "it was difficult to determine that due consideration had been given at the appropriate times to the impact of this series of changes."

### 5.8.2 Adapting to changing process conditions

Kletz provided the following two examples where processes have changed gradually to the point where finally the system becomes hazardous [3].

---

**Text Box 5.16**

These are the most difficult to control. Often, we do not realise that a change is taking place until it is too late. For example, over the years, steam consumption at a plant had gradually fallen. Flows through the mains became too low to prevent condensate from accumulating. On one of the mains, an inaccessible steam trap had been isolated, and the other main had settled slightly. Neither of these mattered when the steam flow was large, but it gradually fell. Condensate accumulated, and finally water hammer fractured the mains.

Oil fields that produce sweet (that is, hydrogen-sulphide-free) oil and gas can gradually become sour. If this is not detected in time, there can be risks to life and unexpected corrosion.

---

### 5.8.3 Gradual change resulting in liquid accumulation in a flare system

A UK gas terminal experienced a very large flame and thick black smoke from its flare system due to accumulation of liquid during plant start-up. [15].

The investigation found that the plant operation had changed gradually over a number of years. This included a change from a continuous to a batch process, decommissioning of equipment, and re-routing process fluids. These changes combined to result in a significant change to plant operation. It was recognised that the site's risk profile had not been reviewed, which meant issues were not properly understood and action was not taken to avoid problems. The company concluded that this was a result of gradual or 'creeping' change occurring because of aging plant. It decided to adopt a methodology for evaluating and managing creeping change in order to avoid similar issues in the future [15].

## 5.9 Modifications to improve the environment

Kletz clearly felt that there were issues with modifications to improve the environment and provided a number of examples [3].

---

### Text Box 5.17

Modifications made to improve the environment have sometimes produced unforeseen hazards. We should, of course, try to improve the environment, but before making any changes we should try to foresee their results.

Explosions in Compressor Houses

A number of compressor houses and other buildings have been destroyed or seriously damaged and the occupants killed, when leaks of flammable gas or vapour exploded. Indoors, a building can be destroyed by the explosion of a few tens of kilograms of flammable gas, but out-doors, several tons or tens of tons are needed. During the 1960s and 1970s, most new compressor houses and many other buildings in which flammable materials were handled were built without walls so that natural ventilation could disperse any leaks that occurred; the walls of many existing buildings were pulled down. In recent years, many closed buildings have again been built to meet new noise regulations. The buildings are usually provided with forced ventilation, but this is much less effective than natural ventilation and is usually designed for the comfort of the operators rather than the dispersion of leaks.

---

Other examples from Kletz included [3]:

---

### Text Box 5.18

In recent years there has been a rapid growth in the number of combined heat and power (CHP) and combined cycle gas turbine (CCGT) plants, driven mainly by gas turbines using natural gas, sometimes with liquid fuel available as standby. Governments have encouraged the construction of these plants, as their efficiency is high and they produce less carbon dioxide than conventional coal and oil-burning power stations. However, they present some hazards, as gas turbines are noisy and are therefore usually enclosed.

In addition, they are usually constructed without isolation valves on the fuel supply lines. As a result, the final connection in the pipework cannot be leak-tested. In practice, it is tested as far as possible at the manufacturer's works but often not leak-tested onsite. Fuel leaks have occurred, including a major explosion at a CCGT plant in England in 1996 due to the explosion of a leak of naphtha from a pipe joint. One man was seriously injured, and a $600\,m^3$ chamber was lifted off its foundations. The precautions that should be taken include selecting a site where noise reduction is not required or can be achieved without enclosure. If enclosure is essential, then a high ventilation rate is needed; it is often designed to keep the turbine cool and is far too low to disperse gas leaks. Care must be taken to avoid stagnant pockets.

Vent Systems

During the 1970s and 1980s, there was increasing pressure to collect the discharges from tank vents, gasoline filling, and the like for destruction or absorption, instead of discharging them into the atmosphere, particularly in areas subject to photochemical smog. A 1976 report said that when gasoline recovery systems were installed in the San Diego area, more than 20 fires occurred in 4 months. In time, the problems were overcome, but it seems that the recovery systems were introduced too quickly and without sufficient testing. As vent collection systems normally contain vapour/air mixtures, they are inherently unsafe. They normally operate outside the flammable range, and precautions are taken to prevent them from entering it, but it is difficult to think of everything that might go wrong. For example, an explosion occurred in a system that collected flammable vapour and air from the vents on a number of tanks and fed the mixture into a furnace. The system was designed to run at 10% of the lower explosion limit, but when the system was isolated in error, the vapour concentration rose. When the flow was restored, a plug of rich gas was fed into the furnace, where it mixed with air and exploded.

In reality improving environmental performance is only one driver of change and there is no fundamental reason why it should be more problematic than any other. However, problems occur because the people who are tasked with implementing the changes do not appreciate the safety risks. Kletz observed how this was starting to improve [3].

## Text Box 5.19

Increasingly, safety, health, and the environment are becoming parts of the same SHE department in industry. This should help to avoid incidents. Unfortunately, there are few signs of a similar integration in government departments.

## 5.10 Managing the risks

Kletz clearly recognised the risks associated with change and his instinct seemed to be that all changes needed to be controlled. This is still recognised today, although the term Management of Change (MoC) is used more often. This change of terminology may have come about as a result of the increased rate of change taking place in industry, driven by technology, meaning that simply trying to control change is insufficient. For most companies there is an almost continual requirement to adopt new technology and so the role of MoC has changed to involve more analysis of how to implement changes effectively without disrupting business.

### 5.10.1 Controlling modifications

Kletz provided very clear advice on how modifications should be controlled [3].

---

**Text Box 5.20**

(1) Before any modification, however inexpensive, temporary or permanent, is made to a plant or process or to a safety procedure, it should be authorised in writing by a process engineer and a maintenance engineer—that is, by professionally qualified staff, usually the first level of professionally qualified staff. Before authorising the modification they should make sure that there will be no unforeseen consequences and that it is in accordance with safety and engineering standards. When modification is complete, they should inspect it to make sure that their intentions have been followed and that it 'looks right'. What does not look right is usually not right and should at least be checked.

(2) The managers and engineers who authorise modifications cannot be expected to stare at a drawing and hope that the consequences will show up. They must be provided with an aid, such as list of questions to be answered.

Large or complex modifications should be subjected to a hazard and operability study.

(3) It is not sufficient to issue instructions about (1) and the aid described in (2). We must convince all concerned, particularly foremen, that they should not carry out unauthorised modifications. This can be done by discussing typical incidents, better still, incidents that have occurred in your own company.

---

This advice is totally sound but not without its problems. Having systems in place that are intended to exert control can act as a barrier. They can make it very difficult and bureaucratic to implement any modifications. This can cause frustration and may lead people to circumvent procedures.

A widely recognised problem is for control procedures to be implemented after it has been decided that the modification will go ahead or even after the modification has been implemented. This puts extreme pressure on people to accept the modification and instead of making an objective assessment they concentrate on developing a justification for why the modification should be authorised.

MoC systems that support the change process instead of simply being focussed on controlling modifications can be more effective because they recognise that change is often required and desirable. They offer far more than a process of deciding if a modification is safe and act as an integral part of the planning, implementation, and review process.

## 5.10.2 MoC system and procedures

A system is required to manage changes. This goes beyond simple control and reflects the stages that need to be followed to drive effective changes whilst managing associated risks.

### 5.10.2.1 Procedures for all types of change

The process for managing change must be fully documented and address the following [16]:

- An explanation of what a modification or change is;
- Assessment by competent people; and
- Assignment of responsibilities to competent people.

As with all procedures it is important to strike a balance between detail and usability. For MoC there is a limit to what can be documented and there will always be a reliance on the people who are actively involved in the process.

Having one procedure to handle all types of change is an option but may be overly complex. Individual procedures for different types of change are another option but can lead to inconsistencies in the way changes are handled.

One effective approach is to have a high level overarching MoC procedure that gives a framework for all types of change. This is then supplemented by specific procedures for different types (e.g. plant, process, procedure, organisation).

Using a range of formats including checklists, decision aids (flow charts), and forms can assist with usability. However, care must be taken to avoid MoC becoming a tick-box exercise or a chase for signatures where the focus becomes more on getting the paperwork straight rather than fully engaging in the important parts of the process.

### 5.10.2.2 Define change

The reality is that almost any intervention can be interpreted as a change at some level. For example, a control room operator entering a set-point into a controller will change how the process is operating but this clearly should not require completion of MoC paperwork provided the set-point is within defined operating parameters.

In practice the requirement is to identify where the MoC process needs to be followed. This depends to a certain extent on the nature of the change. For example with physical changes it is relatively easy to say that MoC applies to any change that is not 'like for like' with a definition that makes it clear that 'like for like' must be the same make, model, and version [5]. On the other hand, any change to the organisation will need to be handled by the MoC [13] because clearly no two people are identical.

Because of this wide-ranging definition it is essential that a filter or other method of classifying changes is provided so that they receive the appropriate level of assessment and control. Applying the same approach to every type of change will inevitably mean that either the system is overloaded with lots of minor changes or that major changes do not receive the attention they require.

### 5.10.2.3 Competent people

Effective MoC is reliant on people at all stages. This includes people with defined roles in the system who will have to assess and approve changes. Also, it includes people working at the 'coalface' (e.g. in the permit to work office, workshops, and stores) [5] who are needed to identify that planned work that has not gone through MoC may constitute a change.

Some of the competencies will be related to the person's 'normal' role and include their technical qualifications and working experience. Others will be related to knowledge of the MoC procedures and the wider issues associated with implementing changes and managing the associated risks.

Different competencies will be required to manage different types of change. Most companies in the process industry will have people with engineering and technical competence to manage plant and process changes. However, finding people with the required mix of human factors and process safety knowledge to assess organisational change can be more of a challenge.

Care must be taken to ensure that the MoC process is not compromised by other project management processes, which tend to be focussed on financial and contractual arrangements.

#### 5.10.2.4 Record keeping

Many incidents have occurred due to people not being aware of changes that have been made in the past. This can occur because changes have been made without following MoC or because the impact of multiple changes over time is not recognised.

Technology can provide solutions such as relational databases providing a mechanism to link changes over time and making the records readily available [14].

### 5.10.3 Implementing change

The purpose of the MoC system is to support the owners of potential changes and the people who have to assess, approve, implement, and review those changes.

#### 5.10.3.1 Justify change

There are two main reasons for change:

1. The change is required; and.
2. An opportunity is identified and making a change is desirable.

Owners of a planned change often perceive that there are no options and hence their role is simply to make sure it happens. They can view MoC as a hurdle that they have to overcome. However, even when changes are required due to obsolescence, changes to regulations, or other external factors, there will usually be options and so part of MoC is making sure that the chosen plan is the best available.

A written statement should be made to clearly define the purpose and benefits of the change. There are three questions which should be answered [16]:

- Is the change necessary?
- What are the benefits?
- Are there any alternatives?

The benefits of the change should be identified as tangible outputs as far as possible. This will allow a worthwhile evaluation after the change has been implemented about whether it was worth doing and/or whether it was implemented effectively. Unfortunately this is rarely done in practice and people often assume that a successful outcome was achieved without any evidence. The types of benefits that should be considered include:

- More efficient process and higher quality outputs;
- Improved reliability;
- Reduced maintenance costs including spare parts;
- Better quality information available allowing better monitoring and control;
- Improved standardisation and connectivity;
- Improved ergonomics and working environment;
- Easier training; and
- Improved security and resilience.

### 5.10.3.2 Understand current arrangements

One of the errors made during MoC is that people select options and plan changes based on an inaccurate understanding of current arrangements. This can mean that the change does not have the expected effect or that the implementation plan does not work in practice and has to be adapted part way through.

Inaccurate knowledge of current arrangements often occurs because people rely on documentation and assume that it is accurate and up to date. Unfortunately it is very common to find that 'as built' drawings are not up to date, often due to unauthorised or poorly documented changes in the past. Also, procedures often do not reflect how a task is performed in practice.

Another error is relying on performance specifications, which are usually what was promised at initial design. In some cases these were never achieved and in others performance may have declined over time. Actual performance can partly be determined from data but consulting people with current experience of operations and maintenance will give a much better picture of what works well and the problems with the current arrangements.

The issues with understanding current arrangements highlight that MoC is not a purely technical activity. Process plants are complex systems consisting of multiple items of hardware, software, and people.

### 5.10.3.3 Describe the new arrangement and transition plan

In order for the proposed change to be assessed and authorised, it needs to be described in enough detail so that people can understand the proposal. This needs to show how the information presented to justify the change will be obtained following implementation, but also how the system will be transformed from its current arrangements.

For smaller modifications it may be sufficient to simply mark-up existing documents to illustrate how the system will be arranged following modification. For larger changes it may be necessary to follow project management principles to assess options and develop an outline design for approval before moving into detailed design.

The implementation plan should identify the activities that need to be carried out, so that the risks during transition can be understood. Again, depending on the scale of the change, this could be a simple list of actions or a more detailed project plan.

### 5.10.3.4 Assess the change and implementation plan

All changes and plans should be assessed before approval. Two aspects of the change need to be assessed [17]:

1. Risks and opportunities resulting from the change (where you want to get to); and.
2. Risks arising from the process of change (how you get there).

The method used should be appropriate for the type and nature of change [5]. HAZOP is not always the answer and other well-known approaches such as Failure Modes and Effects Analysis (FMEA), HAZID, 'What If' analysis, Job Safety Analysis (JSA), or some combination may be appropriate. A site tour and involvement of the people who know the current arrangements and/or may be affected by the change should be considered. The MoC system should include guidelines on determining what type of assessments should be carried out but it must be recognised that technical competence and understanding of the processes used to manage change are essential.

It is important to consult with staff (including contractors) before, during, and after the change [12]. This gives them the opportunity to raise concerns and also makes sure they know what is happening and why.

As always, the requirement should be that the risks are As Low As Reasonably Practicable (ALARP). The justification statement should demonstrate that all options have been considered for both the change itself and the means to achieve it. The change owner should either demonstrate that their chosen solution is the best or present a range of options with the information needed for evaluation.

It is usually wise to avoid too many simultaneous changes. Part of the assessment should be to consider whether it is better to implement the full change in one go or in phases.

### 5.10.3.5 *Implement the plan*

The resources required to implement a change are often underestimated. Sometimes the focus is on the people doing the physical work involved in the implementation without considering the impact on others. For example, a contractor may be brought in to do the work but the operations team will need to prepare plant, issue permits to work, and monitor activities. This becomes even more important if contractors are engaged on a lump-sum basis or the implementation team are given a challenging date for completion as they are incentivised to work quickly, and to ensure quality requires close monitoring that is often not budgeted for.

Resources are often required to prepare for return to service after the change has been implemented. This will include preparing or updating procedures and drawings, and providing training. People doing this will often have to be released from their normal role, which requires others to cover. Also, initial commissioning following implementation may require additional resources to ensure that the people with the appropriate knowledge are present or available in case of problems.

## 5.10.4 Monitor and review implementation

The risks associated with change highlight why it is important to monitor and review what happens in practice. There is no point in assessing changes if they are not implemented as expected or approved.

### 5.10.4.1 *Track progress*

Milestones and targets are inevitable but should not compromise the outcome of the change. If progress is not as expected it is important to understand why. If it is due to lack of resources or changing priorities this is not necessarily an issue with MoC. However, if it is because the implementation is more difficult than expected it can indicate that information used to develop the plan and during assessment and approval of the change was inaccurate. It is much better to recognise fundamental issues early during implementation than to wait until later when changing the plan becomes far more difficult. Unfortunately, problems are often recognised too late and the only option is to continue to completion even though the information available is suggesting that the change is likely to fail.

### 5.10.4.2 *Managing changes during implementation*

There is usually an expectation that every planned change will be successful, and so little thought is given to what could go wrong and the need for a contingency. In reality there will always be a degree of uncertainty as to the impact of change [17]. Signs of unexpected outcomes and risks should be looked for and there should be an acceptance that plans may have to change, including returning the system to its original state (i.e. cancelling the change).

Changes during implementation need to follow the MoC. The end result of the overall change may turn out to be the same as initially intended, but if the way this is achieved is changed this needs to be assessed, documented, and resourced.

### 5.10.4.3 *Review before return to service*

Commissioning plans including pre-start-up safety plans should be in place once the change has been implemented. These should include clear decision criteria about whether the system is fit for commissioning, which will require that all physical work has been completed, documentation updated, and training of personnel complete. It is important to recognise that the effects of change can be subtle or delayed [17].

All paperwork should be up to date including drawings, procedures, asset management, and maintenance systems. There should be a formal approval to start-up.

## 5.10.5 Review the MoC system

Like all management systems, MoC processes should be reviewed on a regular basis. Reviews must consider the effectiveness of MoC in practice and not only on how it is documented. A poorly performing MoC can have major implications for safety.

### 5.10.5.1 *MoC records*

If changes are not properly documented it becomes difficult to keep a view on the as-built arrangements. Unfortunately this is one area where companies often fail. Sometimes this is because the MoC has not been followed correctly so the associated documentation is never generated. In other cases it is simply because documents get lost. Often people are working under pressure and satisfy themselves that the MoC is being followed and that documentation can be dealt with later, but they then move on to the next problem and this does not happen.

The records for each change should include:

- Justification;
- Output from the assessment during the planning stage;
- Progress reports from the implementation stage;
- Results of the pre-start-up review; and
- Feedback following start-up after change implementation.

The records should show that all actions raised by the MoC have been closed effectively. This includes that all drawings, procedures, asset records, and maintenance plans have been updated and are live on the system and previous versions have been archived.

### 5.10.5.2 *Obtain feedback*

Most MoC systems include a requirement for reviews to be carried out after a change has been implemented. This is an important way of making sure that the change has been successful, which should mean that the original justification has been proven and that risks have not increased as a result. Unfortunately this does not happen very often. This can be because the people who were actively involved in the change have moved on to something else, but there is also a behavioural element where people may be reluctant to conduct a review in case the results are less favourable than expected.

The impact of change can be subtle or delayed. This may mean that reliable data is not readily available. However, talking to the people affected can provide a very valuable insight. In fact there have been many changes made that have had no impact simply because they were never used. This can be because the people who needed to use it did not know about the change or understand how they were supposed to use it, they may not have believed it was required or was going to be of any use, or they may have had a bad experience when they did try it so chose not to do it again.

It seems likely that many changes fail to achieve the improvements that were presented to justify them.

### 5.10.5.3 *Ensure MoC is fit for purpose*

It is important that the MoC system is not so cumbersome that good ideas become bogged down in unnecessary or inappropriate reviews. The system itself can become a hazard if safety-related changes stagnate. The solution is to have an efficient filtering mechanism with a competent person in control to ensure that the system is implemented in an effective and efficient way in practice [5].

This is where there is a contrast between control of modifications and managing change. A controlling system that is fit for purpose will generally mean that modifications are avoided unless deemed necessary whereas a MoC that is fit for purpose will provide a practical and effective means of implementation so that a company can take advantage of opportunities whilst managing overall risks.

Maintaining good MoC requires a good system manager who must have clear oversight of the system as a whole and the changes that are taking place [18]. They need to be accountable for how the system functions and ensure that the right people are involved and that the required audits and reviews take place. "The 'good' MoC system that we strive for is not impossible to achieve, but it does take relentless effort from all the stakeholders, all working together towards our ultimate dream ... that perfect Process Safety Management System" [18].

## References

[1] T. Kletz, By Accident—A Life Preventing Them in Industry, PFW Publications, 2000.
[2] T. Kletz, Lessons From Disaster—How Organisations Have No Memory and Accidents Recur, IChemE, 1993.

[3] T. Kletz, What Went Wrong?—Case Histories of Process Plant Disasters and How They Could Have Been Avoided, fifth ed., Elsevier, 2009.

[4] D. Bridger, HF acid leak from tanker unloading flange, Loss Prev. Bull. 232 (2013) 4–9.

[5] C. Feltoe, Recognising small plant changes, Loss Prev. Bull. 239 (2014) 9–11.

[6] Anon, Small or big changes, not managing them can be risky!, Loss Prev. Bull. 267 (2019) 17–20.

[7] R. Stokes, Evangelos Florakis naval base explosion, Loss Prev. Bull. 263 (2018) 9–12.

[8] BBC, Samarco dam failure in Brazil 'caused by design flaws', BBC News (2016), https://www.bbc.co.uk/news/business-37218145 (30 August 2016).

[9] Anon, Anhydrous ammonia leak, Millare, UK, Loss Prev. Bull. 265 (2019) 24–28.

[10] T. Leggett, n.d. What went wrong inside Boeing's cockpit? BBC News, https://www.bbc.co.uk/news/resources/idt-sh/boeing_two_deadly_crashes (Accessed 10 June 2019).

[11] Energy Institute, Managing Major Accident Hazard Risks (People, Plant and Environment) during Organisational Change (2020).

[12] A. Brazier, Organisational change, Loss Prev. Bull. 239 (2014) 3–6.

[13] L. Braben, N. Morris, Organisational Change: Learning From Experience. Hazards 29, IChemE, 2019.

[14] N. Cann, Lessons from a management of change incident, Loss Prev. Bull. 239 (2014) 12–14.

[15] R. Berriman, Creeping Change—Liquid Accumulation in a Flare System at a Gas Terminal. Hazards 29, IChemE, 2019.

[16] G. Gill, Controlling plant modifications, Loss Prev. Bull. 239 (2014) 17–20.

[17] Health and Safety Executive, HSE Information Sheet CHIS7—Organisational Change and Major Accident Hazards. HSE Information Sheet, (2003).

[18] K. Patterson, G. Wigham, Management of change—what does a 'good' system look like? Loss Prev. Bull. 267 (2019) 7–9.

# Human error

## 6.1 Introduction

According to his autobiography [1] Kletz identified that a new attitude to human error was required as the result of a 'realisation that most accidents could be prevented (or made less likely) by managers' actions. To say accidents are due to human failing is true but is as helpful as saying falls are due to gravity.

This is another area where Kletz was ahead of his time, publishing his thoughts on the subject in the early 1980s. At this time the only other people with an interest in the subject were psychologists working in other industries such as nuclear and aviation. The first edition of his book An Engineer's View of Human Error was published in 1985.

Kletz described how he came to realise the importance of, what we would call today, human factors [2].

### Text Box 6.1

Many years ago, when I was a manager, not a safety adviser, I looked through a bunch of accident reports and realised that most of the accidents could be prevented by better management - sometimes by better design or method of working, sometimes by better training or instructions, sometimes by better enforcement of the instructions.

Together these may be called changing the work situation. There was, of course, an element of human failing in the accidents. They would not have occurred if someone had not forgotten to close a valve, looked where he was going, not taken a short-cut. But what chance do we have, without management action of some sort, of persuading people not to do these things?

To say that accidents are due to human failing is not so much untrue as unhelpful, for three reasons:

(1) Every accident is due to human error: someone, usually a manager, has to decide what to do; someone, usually a designer, has to decide how to do it; someone, usually an operator, has to do it. All of them can make errors but the operator is at the end of the chain and often gets all the blame. We should consider the people who have opportunities to prevent accidents by changing objectives and methods as well as those who actually carry out operations.

**(2)** Saying an accident is due to human failing is about as helpful as saying that a fall is due to gravity. It is true but it does not lead to constructive action. Instead it merely tempts us to tell someone to be more careful. But no-one is deliberately careless; telling people to take more care will not prevent an accident happening again. We should look for changes in design or methods of working that can prevent the accident happening again.

**(3)** The phrase 'human error' lumps together different sorts of failure that require different quite actions to prevent them happening again.

This chapter will look at what Kletz said about the errors made by different groups of people and the solutions he proposed for reducing risks. It will update the text with some new examples and latest thinking on the subject.

## 6.2 Different types of error

Kletz used the following spelling examples to illustrate that there are different types of human error and how each requires a different solution [2].

---

**Text Box 6.2**

As an example of the five sorts of error, consider spelling errors:

- If I type opne or thsi it is probably a slip, overlooked when I checked my typing. Telling me to be more careful will not prevent these errors. Using the spell-checking tool on my word processor will prevent many of them.
- If I type recieve or seperate, it is probably due to ignorance. If I type wieght it may be the result of following a rule that is wrong: 'I before e. except after c'. Better training or instructions might prevent the errors but persuading me to use the spell-checking tool would be cheaper and more effective.
- If I write Llanfairpwllgityngyllgogerychwyrndrobwlllantysillogogogoch (a village in Wales), it is probably because it is beyond my mental capacity to spell it correctly from memory. (Can you spot the error?)
- If I type thru or grey, it is probably due to a deliberate decision to use the American spelling. To prevent the errors someone will have to persuade me to use the English spelling.
- If I work for a company, perhaps the managers take little or no interest in the standard of spelling. They may urge me to do better but do not realise that they could do more, such as encouraging me to use the spell-checking tool.

---

This illustrates why we need to recognise that different solutions are required depending on what people are doing and the types of errors they may make. This is the crux of human factors, which is concerned with reducing the risks that occur due to human error.

Whilst we are all different we all make similar types of errors. The consequences of our errors will depend on our job and the task we are performing at the time, but human factors can be applied in all cases. When an accident occurs it is often easy to identify the errors made by the people present at the time, usually the operators or maintenance technicians, because they will have been the last people to have been actively involved with the system. But many other people will have been involved in the proceeding hours, days, or years and their contribution to the eventual accident is often far greater than the errors that occur in the minutes or seconds immediately before the accident. Kletz was very well aware of this and made it clear that we need to understand it if we want to really to improve the way we manage risks and prevent accidents.

## 6.2.1 Operator errors

Operators interact with systems when they are in their most hazardous state (i.e. the system is most likely to contain hazardous materials and/or conditions such as pressure and temperature). If operators make errors that result in loss of control of the hazard the subsequent accidents can have serious consequences. It is right to view the management of operator error as a priority, but it must be recognised that operators' errors often occur due to errors made by other people (e.g. designers, managers) or are triggered by technical failures.

The following is an example of an operator error described by Kletz [2].

---

**Text Box 6.3**

A suspended catalyst was removed from a process stream in a pressure filter (Fig. 6.1). When a batch had been filtered, the inlet valve was closed and the liquid in the filter blown out with steam. The steam supply was then isolated, the pressure blown off through the vent and the fall in pressure observed on a pressure gauge. The operator then opened the filter for cleaning. The filter door was

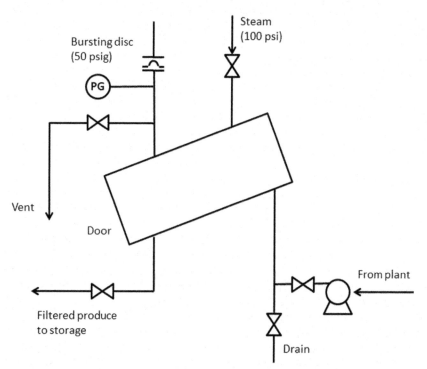

FIG. 6.1    Filter arrangement [2].

held closed by eight radial bars which fitted into U-bolts on the filter body. To withdraw the radial bars from the U-bolts and open the door the operator had to turn a large wheel, fixed to the door. The door, with filter leaves attached, could then be withdrawn.

One day an operator, a conscientious man of great experience, started to open the door before blowing off the pressure. He was standing in front of it and was crushed between the door and part of the structure. He was killed instantly.

This accident occurred some years ago and at the time it seemed reasonable to say that the accident was due to an error by the operator. It showed the need, it was said, for other operators to remain alert and to follow the operating instructions exactly. Only minor changes were made to the design.

However, we now see that in the situation described it is inevitable, sooner or later, that an operator will forget that he has not opened the vent valve and will try to open the filter while it is still under pressure. The accident was the result of the work situation, and we would recommend changes in the design.

---

This was a very simple error of forgetting a step in a task. Everyone makes these types of error but when an operator does they are often exposed to a hazardous situation.

Kletz highlighted how the types of accidents involving Operator errors can vary significantly, for example [2]:

- Restarting a reactor stirrer that had stopped mid-batch without realising the potential for a violent reaction when reactants are mixed suddenly;
- Damaging equipment when using pressure to clear blocked pipework without realising the potential for stored energy;
- Overfilling road and rail tankers because of inattention, setting the wrong quantity on a delivery meter, or failing to take account of product already on board;
- Over-pressurising pressure vessels by hydraulically filling and not allowing for thermal expansion.

### 6.2.1.1 Intuitive design

We can give operators more education so that they understand the hazards they are dealing with and what they need to do to keep safe. But operators perform many routine tasks where the natural human reaction is to start working on 'auto pilot' instead of paying full attention. This is not because they are bad people or operators; they are only human.

Back in the 18th century, English poet Alexander Pope wrote "To err is human" which continues to be one of the most popular truisms of the past 200 years. The fact that people make mistakes is not a surprise to most people, and this is at the heart of intuitive design. Intuitive design recognises the tendency of people to make mistakes is a natural human condition that we need to take into account in the way we design systems.

Intuitive design results in systems that are easy to use without having to consciously think about what is being done and it is the most effective way of reducing routine errors.

Consistency is particularly important. If all local control panels on a site are arranged with the start button at the top and stop button at the bottom, the operator will instinctively know

FIG. 6.2    Inconsistent layout means operators need to think each time they use the buttons. (All of these are equally acceptable in isolation.)

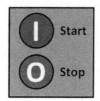  

FIG. 6.3    Mirror images create confusion.

which button and is far more likely to choose the right one every time. If there is no consistency they are required to pay more attention to make sure they select the correct button and are more likely to select the wrong one from time to time Fig. 6.2.

Similarly, equipment designed with a mirrored layout (e.g. equipment panels, adjacent spared pumps) are many times more likely to result in a mistake in operation than those designed with the same layout, because they are not intuitive—they don't put the operator at the centre of the design process and do not set the operator up for success Fig. 6.3.

Avoiding laying error traps for unsuspecting operators is a focus of intuitive design.

### 6.2.1.2 Error recovery

Unfortunately, simple errors made by competent people when performing routine tasks cannot be prevented entirely. Another risk control strategy is to give people the opportunity to recover from their errors before a consequence occurs. However, this is not without its problems as shown in Table 6.1.

This highlights that there is no perfect solution and eliminating human error completely is not possible. Human factors assist us in achieving risks that are As Low As Reasonably Practicable (ALARP) whilst acknowledging the risk can never be zero.

### 6.2.1.3 Formosa accident

On April 23, 2004, five workers were killed and two others were seriously injured when an explosion occurred in a polyvinyl chloride (PVC) production unit at Formosa Plastics in Illiopolis, Illinois, east of Springfield. The explosion followed a release of highly flammable vinyl chloride, which ignited. The explosion forced a community evacuation and started fires that burned for several days at the plant [5]. The immediate cause was an operator opening a

TABLE 6.1    Potential pitfalls with error recovery strategies.

| Possible error recovery strategy | Potential pitfalls |
|---|---|
| Confirmation 'are you sure' messages allowing the operator to cancel the action | People very quickly learn to expect these messages and accept them without reading what they say |
| Warning alarm | May contribute to alarm overload, which is a known problem for many organisations [3] |
| Interlock to prevent a step being performed when conditions are not correct | May result in overreliance, assuming the possible errors are prevented without recognising that interlocks are not 100% reliable [4] |

drain valve on the wrong reactor resulting in a release of vinyl chloride. He did this because of a very simple mistake. He turned the wrong way after descending a flight of stairs, failing to recognise his error because all the reactors looked very similar.

The drain valve had been fitted with a safety interlock to prevent it being opened when the reactor was pressurised. However, this was easily defeated by applying air from a hose that had been provided in the area because in certain emergencies it would be necessary to open the valve to drain when the reactor was pressurised. In this case it was concluded that the operator thought he was at the correct reactor and suspected a fault with the drain valve mechanism was stopping it from opening. It seems very likely that he convinced himself that his action to defeat the interlock was justified and perfectly safe.

In this incident, the design was not intuitive and there was no easy way available to the operator to recognise a mistake had been made and to recover from the error.

This accident not only highlights how operators can make simple errors (going to the wrong reactor), but also that their drive to get the job done can lead them to make decisions, which in hindsight are inappropriate. This means measures made to reduce the likelihood of operator error have to be particularly robust. In this case, eliminating the need to routinely drain reactors to an open drain system may not have prevented the operator error but could have prevented the accident. Also, designers probably felt that the interlock on the drain valve had eliminated the possibility of operator error when in fact it had, at best, only reduced the likelihood, which was not sufficient given the potential consequences as illustrated by this accident.

### 6.2.1.4 Understanding operators' actions

Kletz described an accident where two men were killed by a boiler explosion [2]. The operator had either not noticed that the water level was dangerously low or did not understand its importance. Designers had recognised the risk of low water level and had provided a range of protective devices for this scenario. However, these were all isolated at the time of the accident (see Chapter 5 for more details of this accident). Kletz made the point that the reason why the operator continued to operate was unknown but it is not unusual for this to occur. The Chernobyl nuclear power plant disaster illustrates how operators can fall into this trap because they feel the circumstances justify the action (see accident summaries). In that case they overrode a number of safety features because they thought it was required to run the test they have been instructed to perform. If we are going to provide effective mechanisms to manage the risks of error we need to understand how operators act in the real world and the factors that influence their performance.

Operator errors must always be a concern for any organisation because they are most likely to cause the greatest harm to people, the environment, equipment, and business performance. Intuitive design of plant, equipment, systems, and procedures has to be primary method of reducing the risks.

## 6.2.2 Maintenance errors

People carrying out maintenance activities can make errors in very similar ways to operators. The difference is that they are usually working on a system that has been made safe in advance so their errors do not immediately result in an accident.

The main concerns for maintenance errors include:

- An accident occurring whilst performing the maintenance because of a failure to control residual hazards;
- An accident occurring due to working on the wrong system;
- An accident occurring when hazards are reintroduced during return to service;
- Latent failures being introduced resulting in failure sometime later after return to service;

Failure to carry out required preventative or reactive maintenance leading to system failure may also be described as a maintenance error, but is generally a result of management errors.

### 6.2.2.1 Examples of maintenance errors

Maintenance errors have been discussed in detail in Chapter 5. But the following examples of Kletz's illustrate that there are many different types of maintenance error [2]:

- Removing the whole thermowell resulting in a loss of containment instead of removing the temperature measuring device;
- Dismantling the pressure-containing elements of a valve instead of removing only the actuator;
- Fitting a filter rated for the wrong pressure due to an error on a drawing that was not recognised by the maintenance technician;
- Undoing all the bolts of a joint allowing trapped pressure to be released instead of slackening one side of the joint and driving a wedge to release the pressure in a controlled way;
- Failure to make up flameproof electrical enclosures correctly so that they would not be gas tight and could result in ignition of a flammable gas cloud;
- Using out of date lifting equipment;
- Failing to obtain a permit to work or deviating from the permitted activities.

### 6.2.2.2 Maintaining the wrong equipment

Maintaining the wrong equipment is a serious concern for any organisation, particularly where maintenance is carried out whilst some parts of the system remain live and where multiple items of similar or identical plant are used. Signage and labelling is particularly critical, but unlikely to be sufficient. Planning for maintenance should be based on the assumption that identification errors will be made.

Incorrect labelling can be a contributory factor to incidents. In one incident, a sign which identified a decommissioned line was actually located nearer the line which was live, rather than the decommissioned line, which years later led to a maintenance technician hooking up to the wrong equipment.

Some companies have implemented procedures with additional controls for the first break of containment into a system. This typically requires a competent operator to be present with the maintenance team to ensure positive identification and to carry out tests to confirm the status of the plant (i.e. confirm it is not still under pressure or full of fluid). This can be effective, although time consuming, provided the procedure is followed in practice.

### 6.2.2.3 Straying beyond scope

Maintenance personnel can be particularly prone (typically more than people in other roles) to do more than was initially planned for a piece of work. This can be where they see an opportunity to reduce downtime and maintenance costs by doing additional tasks or where they follow an alternative method when they encounter problems or unexpected situations. This can be perpetuated by a view that when the plant has been shutdown for one piece of maintenance it is safe to do almost anything.

Effective permit to work systems supported by good communication are critical to making sure only planned and agreed work is carried out and that any deviation requires further approval.

### 6.2.2.4 Accident at Pont de Buis, Finistère, France

On July 2014 a pyrotechnics manufacturing site in Pont de Buis, Finistère, France experienced an explosion during maintenance [6]. Three technicians were seriously hurt when they saw ignited pyrotechnic dust that had accumulated in pipework that was being cut. The task being performed as part of a three-week annual maintenance shutdown was to replace and disassemble a dye tank. Whilst doing this the technicians had taken the initiative to remove an additional section of pipe that was not covered by the plan. The individuals involved were considered to be very experienced and overconfidence may have prompted them to stray from the plan. Lack of supervision during the annual relief maintenance period and inadequate training in the recognition of potential hazards were identified as the likely root causes.

### 6.2.2.5 Errors when maintaining safety critical devices

The potentially most significant errors are those made during maintenance of safety critical devices including pressure relief devices and Safety Instrumented Systems (SIS). Maintenance of these can be more complicated than it first appears and it is easy for people to think they are performing them correctly without realising that many different errors are possible. An additional problem is that these errors are very difficult to detect and may only be revealed when these devices, which we typically rely on in the most hazardous situations, fail.

Failure of Bursting Disks (BD) or Pressure Safety Valves (PSV) to operate on demand can be catastrophic. To reduce the likelihood of this they are routinely replaced (BDs and some PSVs) or removed for calibration and servicing before refitting (PSV). The problem is that these essential activities introduce potential for error including:

- Selecting a BD rated for the wrong pressure or calibrating PSV incorrectly;
- Mishandling during transport or in storage so that they get damaged or deteriorate;
- Failing to detect debris (including packaging material) and contamination in pipework that can lead to blockages in the inlet or outlet, affecting the ability of the device to relieve pressure when required;
- Installation errors that can affect operation;
- Leaving the device isolated.

People often focus on the physical aspects of these maintenance tasks that can have immediate consequence (i.e. breaking and making joints) and overlook the more subtle but potentially critical errors that can stop the device operating on demand.

Similar issues arise with SIS and the proof testing that is carried out routinely to provide reassurance that they will operate when required. Potential maintenance errors include:

- Assuming a test of individual components or parts of systems gives an accurate assessment of the whole system;
- Exercising components before the proof test so that the results achieved do not reflect how they would operate if there is a demand during operation (e.g. operating a valve before the test to free it up so that it closes quicker than it would have before being operated);
- Failure to test every possible unrevealed failure;
- Failure to test every combination of initiator operations, particularly for 'voted' systems (e.g. two out of three transmitter activation);
- Leaving the system or components isolated or over-ridden.

People do not recognise these potential errors and or that the methods they have been using for proof testing are providing unreliable results.

#### 6.2.2.6 Inadequate record keeping

There can be a tendency to view that returning a system to its required state is the only critical objective of maintenance. If records are not kept about reactive maintenance that is carried out or the findings from preventative maintenance of more fundamental issues (e.g. weaknesses in system design) may not be recognised.

This applies to all items including safety critical devices (e.g. BD, PSV, and SIS). Recurrence of similar faults and problems is a very important indicator of overall risk but is unlikely to be recognised if records are not kept of maintenance activities or the records are not routinely analysed. Maintenance personnel need to be aware of the need to record everything they do in a form and given a mechanism for doing this that will facilitate analysis.

### 6.2.3 Design errors

Kletz's view that many operator and maintenance errors are actually caused by design errors has now been widely accepted. There are two types of error made by designers:

1. Technical errors that mean the system they design is unable to meet its objectives in terms of process or safety;
2. Non-technical errors that mean the system is difficult to operate or maintain.

The following describes Kletz's view on the first type of design error [2].

---

### Text Box 6.4

I do not, of course, wish to imply that accidents are due to the negligence of designers. Designers make errors for all the reasons that other people make errors: a lapse of attention, ignorance, lack of ability, a deliberate decision not to follow a code. Unlike operators they usually have time to check their work, so slips and lapses of attention are frequently detected before the design is complete. Just as we try to prevent some accidents by changing the work situation, so we should try to prevent other accidents by changing the design situation—that is, we should try to find ways of changing the design situation so as to produce better designs. The changes necessary will be clearer when we have looked at some examples, but the main points that come out are:

- Cover important safety points in standards or codes of practice.
- Make designers aware of the reasons for these safety points by telling them about accidents that have occurred because they were ignored. As with operating staff, discussion is better than writing or lecturing.
- Carry out hazard and operability studies on the designs. As well as the normal HAZOP on the line diagrams, an earlier HAZOP on the flowsheet (or another earlier study of the design concept) may allow designers to avoid hazards by a change in design instead of controlling them by adding on protective equipment
- Make designers more aware of the design concepts.

---

The potential for design errors must be reflected in the procedures followed by designers and arrangements for supervision. Also, the way designs are reviewed.

#### 6.2.3.1 Flash fire at the US ink/sun chemical corporation

On October 9, 2012 a flash fire in a newly installed dust collection systems caused burn injuries to seven workers at the US Ink/Sun Chemical Corporation ink manufacturing facility in East Rutherford, New Jersey, USA [7]. The investigation concluded that the design of dust pick-up points resulted in excessive quantities of condensable vapours being extracted. These vapours condensed to form sludge in the ductwork causing it to block. This led to low conveying velocity and an accumulation of dust, which was the fuel for the primary deflagration that initiated the incident chain of events. Also, the design did not include sufficient provision to prevent, contain, or extinguish fires.

Whilst the investigation report demonstrated that the designers of the dust collection made a number of errors it also highlighted a lack of management oversight in the planning, design, installation, and commissioning. There was a failure to follow company procedures requiring a Process Hazard Analysis (PHA) to be carried out for this new installation. Investigators concluded that a PHA would have triggered consideration of additional safety factors.

### 6.2.3.2 Safety studies

One development in recent years has been the acceptance that safety studies (e.g. HAZOP, LOPA, Bow-tie diagrams) should be considered as an integral part of projects. There is rarely any prescriptive legislation that mandates particular studies, but certain studies (particularly HAZOP) have been adopted widely as a project requirement.

There is usually benefit in carrying out a number of different studies because they each give an opportunity to look at the design from a different perspective. Formally considering human factors is a fairly recent addition to projects. One question has been whether this should be addressed as part of other studies or whether additional studies are required. There have been some suggestions that studies can be combined or consolidated, but this reduces the opportunities to consider a design from different perspectives.

### 6.2.3.3 Human factors engineering (HFE)

A recent addition to the standard safety studies is Human Factors Engineering (HFE). It is a multidisciplinary science that is concerned with understanding interactions between people and systems. It applies scientific knowledge and principles to ensure systems are designed in a way that optimises the human contribution to production and minimises risks to health, personal or process safety and environmental performance [8]. It can be very effective at identifying the non-technical errors made by designers.

The typical aims of HFE are to ensure design reduces health, safety, and environmental risks by eliminating or reducing the likelihood of human error. Additional aims are to improve human efficiency and productivity in order to enhance operational performance and to improve user acceptance of new facilities.

An HFE review should take place in the early stages of a project [9] to identify the human factors risks and to develop a plan to address them through design. A key to success is to involve people with operational experience in the project as this has been found to be very successful at creating systems that match the requirements of operators and maintenance personnel.

Projects should also create HFE plans [8] which list the activities to be undertaken in the lifecycle of the project, such as:

- Defining project philosophies for accessibility, automation, staffing arrangements, Human Machine Interface (HMI) design and alarms;
- Task Requirements Analysis (TRA) to define general requirements and Safety Critical Task Analysis (SCTA) to define specific requirements to be incorporated in the design;
- HFE design reviews to confirm the requirements identified have been addressed;
- Staffing and workload analyses;
- Alarm review and rationalisation;
- HFE ALARP demonstration.

### 6.2.3.4 Ensuring accessibility

Kletz's following example [2] illustrates that design is often a compromise, particularly where human factors are concerned.

**Text Box 6.5**

Operators often complain that valves are inaccessible. Emergency valves should, of course, always be readily accessible. But other valves, if they have to be operated, say, once per year or less often, can be out of reach. It is reasonable to expect operators to fetch a ladder or scramble into a pipe trench on rare occasions.

Designers should remember that if a valve is just within reach of an average person then half the population cannot reach it. They should design so that 95% (say) of the population can reach it.

HFE provides a means for achieving a reasonable compromise in an objective way. The general philosophy should be that all items requiring manual operation or maintenance (e.g. valves and instruments) are accessible and/or visible from ground level or a platform, at a height that is within a person's reach and visibility. Where access is required particularly frequently or may be required urgently (e.g. in an emergency) the access should be more clearly defined so that it is within an ideal height. Conversely, where access is required very infrequently there is no requirement to provide permanent access and temporary scaffold or Mobile Elevating Work Platform (MEWP) is acceptable.

This approach allows designers to make sensible decisions and provide objectivity in design reviews. It has been formalised in projects as a Valve and Instrument Criticality Analysis (VICA).

### 6.2.3.5 *Allocation of function*

Allocation of Function has been provided by the human factors community as a method of deciding whether a particular function is best suited to being performed by a person, by technology (hardware or software), or a mixture of both. The following example illustrates Kletz understood the need to design for human capabilities, even though he may not have been aware of the term allocation of function [2].

**Text Box 6.6**

One design engineer, finding it difficult to install the devices recommended (quick release closures with interlocks on pressure containing equipment), said it was 'reasonable to rely on the operator'. He would not have said it was reasonable to rely on the operator if a tonne weight had to be lifted; he would have installed mechanical aids. Similarly if memory tasks are too difficult we should install mechanical (or procedural) aids.

Of course, we can rely on the operator to open a valve 99 times out of 100, perhaps more, perhaps less if stress and distraction are high, but one failure in 100 or even in 1000 is far too high when we are dealing with an operation which is carried out every day and where failure can have serious results.

Many design engineers accept the arguments of this book in principle but when safety by design becomes difficult they relapse into saying, 'we shall have to rely on the operator'. They should first ask what failure rate is likely and whether that rate is tolerable.

Kletz's previous example highlighted a simple consideration of operator capability. Allocation of function applies to all activities, including far more complex operations and procedures, and should be a key consideration when introducing automation.

With advances in technology there has been a tendency to automate everything that is easy to automate with the justification that it will eliminate the risk of operator error. However, automation is never 100% reliable and there are potential pitfalls that are often overlooked by designers including:

- Rather than risks being eliminated, the potential human errors are transferred from the operators to the maintenance personnel who have to ensure the automation is working as intended;
- Operators have less opportunity to operate the system on a day-to-day basis leading to a loss of competence and an increased likelihood of human errors when the automation fails;
- Reduced understanding of how the system works making it more difficult to evaluate risks including the potential consequence of human errors when operating or maintaining the system.

Designers should be specifying automated systems based on a realistic view of the human and technical risks. Part of this should be a means for maintaining operator competence. One advantage of the advances in technology is that simulators can be obtained relatively cheaply and these can provide operators the opportunity to practise dealing with infrequent and complex events. This has been common in some industries (e.g. airline pilots) for many years but the take up in the process industries has been fairly slow.

### 6.2.3.6 Use of design contractors

Kletz highlighted a problem with the use of design contractors that remains to this day [2].

---

**Text Box 6.7**

Design contractors who are asked for a cheap deign are in a quandary. Do they leave out safety features and safety studies such as HAZOP, which are desirable but perhaps not essential, in order to get the contract? Obviously there is a limit beyond which no reputable contractor will go but with this reservation perhaps they should offer a minimum design and then say that they recommend that certain extra features and studies are added.

---

Lump sum contracts awarded using competitive tendering can increase the risk that short cuts are taken. This is arguably greater for Engineering, Procurement, and Construction (EPC) contracts because designers are very well aware that the decisions they make can have a significant impact on the project cost and their employer's profit.

Clients try to minimise project risks by listing many different standards that the designer has to comply with. The problem with this is that the designer is presented with an overwhelming number of requirements that can be difficult to understand and sometimes contradictory. Detailed, project specifications can be more effective but expensive to develop and can narrow options too early in a project so that opportunities for improvements in the design are missed.

In an ideal world contracts would result in an open and sharing relationship between design contractor and client. But in the real world, at least understanding the issues and how they can influence the designers' behaviour can at least allow controls to be put in place.

### 6.2.3.7 Designers' experience

People in design teams often have little or no operations experience. This can be the case for in-house teams, but is usually worse where design contractors are used. For this reason it is essential that people with relevant operations experience take an active role in any project.

One of the frustrations for design teams is that the right people are not released to attend meetings, etc. With tight timescales and fixed budgets they cannot wait for people to become available and so the design moves on without obtaining the right input. It is important that operations input is not only identified as essential but that someone is given the responsibility to make it happen, with that person having the authority to release Operators to attend where required.

It is very rare for maintenance personnel to be involved in projects. The result is that insufficient thought is given to how a new plant will be maintained.

## 6.2.4 Construction errors

Errors made by construction personnel can cause immediate problems during commissioning and longer term issues if they introduce weaknesses or other latent failures. There are three types of error made by construction personnel:

1. Misinterpreting design information so that the construction is not as intended;
2. Operational errors made by construction personnel;
3. Making decisions during construction regarding details that are not fully specified in design that make the system more difficult to operate or maintain.

Some construction errors can be detected during site inspections and pre-start-up safety reviews. However, many will not be immediately visible or detectable.

### 6.2.4.1 Straying from design intent

Well intentioned decisions made during construction can result in a departure from the original design. This can occur because they are put under pressure to cut costs or complete more quickly, but often the root cause is that they do not understand the design objectives. Kletz used the following example to illustrate how contractors can stray from design intentions during construction [2].

---

**Text Box 6.8**

A compressor house was designed so that the walls would blow off if an explosion occurred inside. The walls were to be made from lightweight panels secured by pop rivets.

The construction team decided that this was a poor method of securing wall panels and used screws instead. When an explosion occurred in the building the pressure rose to a much higher level than intended before the walls blew off, and damage was greater than it need have been. In this case a construction engineer, not just a construction worker, failed to follow the design. He did not understand, and had probably not been told, the reason for the unusual design.

In this case the contractors had a legitimate concern regarding the method proposed for securing the wall panels. In some instances they may have been congratulated for using their initiative but in a major hazard setting any change can be critical and should be reviewed and authorised.

It is impossible to check every construction detail. Construction personnel must understand the boundaries that they have to work within and adhere to a change management process (there can be a tendency to avoid reporting issues in case it delays the project). Equally, construction personnel should be encouraged to raise concerns if they identify issues with the design or have suggestions for improvement.

### 6.2.4.2 Design constructability

Kletz's following example highlights that the root cause of many construction errors is in design [2].

---

**Text Box 6.9**

A fire at a refinery was caused by corrosion of an oil pipeline just after the point at which water had been injected (Fig. 6.4A). A better design is shown in Fig. 6.4B. The water is added to the centre of the oil cream through a nozzle so that it is immediately dispersed. However a plant that decided to use this system found that corrosion got worse instead of better. The nozzle had been installed pointing upstream instead of downstream (Fig. 6.4C).

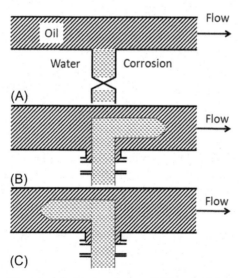

FIG. 6.4    (A) Original design, (B) revised design—as designed, (C) revised design—as fitted [2].

---

Ease of correct construction should be a key objective of design and potential construction errors should be identified so that the risks can be managed by providing better instructions and scheduling appropriate checks and reviews. In Kletz's example the problem was that it was not possible to check the nozzle orientation after it had been installed.

Improved constructability can have knock-on benefits with future maintenance.

### 6.2.4.3 HFE plan covering construction activities

One output from the HFE process should be a plan to ensure human factors are properly considered during construction. This can address the operational errors by construction personnel but is mainly focussed on ensuring that the detailed decisions made during construction make a positive contribution to reducing operator and maintenance errors.

Kletz highlighted the importance of communicating requirements to construction personnel, which is something a good HFE can assist with [10].

---

**Text Box 6.10**

Errors by construction teams are best detected by detailed inspection during and after construction. Who should carry out the inspection? The checks made by construction inspectors in the past are clearly not sufficient.

The team who will start up and operate the plant have an incentive to inspect the plant thoroughly, as they will suffer the results of any faults not found in time. The designers, though they may not spot their own errors, may see more readily than anyone else when their intentions have not been followed. The inspection of the plant during and after construction should therefore be carried out by the start-up team assisted by one or more members of the design team.

Could we reduce the number of construction errors by taking more trouble to explain to construction workers the nature of the materials to be handled and the consequences of not following the design and good practice? Usually little or no attempt is made to carry out such training and many construction engineers are sceptical of its value, because of the itinerant nature of the workforce. Nevertheless, perhaps it might be tried. Certainly, there is no excuse for not telling construction engineers and supervisors why particular designs have been chosen, and thus avoiding errors such as some of those described.

---

A key part of the HFE plan should be how features that may not be fully detailed in the design should be handled. Examples include routing of cables and small bore pipework, orientation of gauges, local control panels, lighting units, signs and labels, arrangement of insulation, and painting (colour coding). These are important because they can have a significant impact on the way operations and maintenance tasks are performed if they form obstacles or confusion.

Construction personnel often have no operations or maintenance experience and so may not recognise potential issues or be able to choose optimum solutions. This means people with appropriate experience have to be readily available to provide advice, monitor activities, and approve decisions.

The HFE plan should clearly identify responsibilities and identify people with appropriate levels of human factors competence to fulfil these roles for the Client, Design, and Construction teams. Focal points with awareness of HFE should be appointed for every discipline (e.g. mechanical, electrical, civil) and they should be supported by people with a higher level of HFE competence.

Communication arrangements should ensure that issues are raised immediately and escalated to responsible parties as required. In addition, planned interventions should include:

- Regular meetings (could be weekly) to discuss issues and learnings from the project;
- Regular inspections of work in progress;
- Formal reviews at key stages in the construction, focussing on any HFE critical items identified during design;
- Construction completion verification;
- Audits focussed on the effectiveness of HFE plan for construction and to identify if the plan needs to be improved and to capture learning for future projects.

### 6.2.5 Programming errors

Kletz was very well aware of the potential issues with increased use of computers with operational functions and included a chapter titled 'Problems with Computer Control' in both his book on Human Error [2] and What Went Wrong? [10].

#### 6.2.5.1 *Ammonia release at Millard refrigerated services, Alabama*

23rd August 2010 a significant quantity of ammonia was lost at the Millard Refrigerated Services facility in Theodore, Alabama when hydraulic shock caused pipework to fail releasing 14.5 tonnes of anhydrous ammonia [11]. Over 800 people were exposed with over 150 workers were immediately affected. Thirty-two people were admitted to the local hospital suffering from the effects of serious exposure and four of these were treated in the intensive care unit.

The incident occurred when the plant was being returned to service following a power failure to the site. Manual interventions carried out by the operators to reset the control system led to the process control computer interrupting the defrost cycle in the chiller, switching the plant directly back to refrigeration mode without first removing the hot ammonia gas from the coils.

The investigation concluded that "the control system contained a programming error that permitted the system to go from soft gas directly to refrigeration mode without bleeding the high pressure from the coil or preventing the low-temperature suction valve from opening. The error with the software logic in the control system went undetected because under normal operations, in its programmed sequence, the defrost cycle would not allow the ammonia liquid to enter the evaporator until the coil was properly depressurized via the bleed cycle."

### 6.2.5.2 *Consequences of programming errors*

Programming errors typically result in either software stopping all outputs in certain situations or active failures where software gives an incorrect output. These can occur because the programmer simply makes a mistake in the code or is working with an incorrect or incomplete understanding of how their software will be used in practice. Typical issues that cause problems include:

- Failing to consider every potential mode of operation including planned and unplanned events;
- Being unable to handle mathematically impossible equations (e.g. a zero input being used as the divisor);
- Failing to recognise data can be presented in different formats (e.g. date formats such as day:month:year vs year:month:day);
- Developing code that requires more resource than may be available some or all of the time.

For any complex systems programming errors should be considered as inevitable and plans should be in place to manage the risks. This is a demanding and specialised subject. Verifying software for correctness is often impossible. Good systems have to be in place to manage programming and to ensure programmers follow good practice.

### 6.2.5.3 *Unexpected output*

A failure to dose a chemical stream was discovered by operators more than 24h after it had stopped. It occurred after an update to the plant's control system software. The operators had not realised that the system was programmed to put the dosing mechanism into manual control with zero flow when it was being updated. Also, they did not realise that the control system showed the last known healthy signal from the dosing flow meter, which was not the actual value if the signal was considered to be unhealthy. The operators saw a value on the control system readout, which gave them a strong but wrong indication that chemical dosing was taking place.

### 6.2.5.4 *Modular programming*

Breaking a system down into smaller modules is one way of managing the risk because an error in one is localised and can be contained. The way these modules interact is important and potential for errors should be taken into account so that they can be handled appropriately.

Complexity of code can have significant long-term effects. It makes it particularly difficult to check the code and to manage future changes. Usually software used to operate process plants is bought as part of control and safety systems. Companies need to be aware of the potential risks of programming errors and factor this into their specifications, vendor selection, and acceptance testing processes.

## 6.2.6 Management errors

Kletz was well aware that most of the errors described in this chapter are caused or contributed to by errors made by managers. His view was that managers make errors due to lack of training or ability. The following explains how Kletz saw this happening [2].

**Text Box 6.11**

If output, costs, efficiency or product quality require attention, senior managers identify the problems, agree actions and ask for regular reports on progress. This approach is rarely seen where safety is concerned. Although senior managers repeatedly say that safety is important, they rarely show the level of detailed interest that they devote to other problems areas. Many only look at the lost-time accident rate but in many companies today all it measures is luck and the willingness of injured employees to return to work. If senior managers comment on the lost-time accident rate and nothing else they give staff the impression that they are not interested in the real problems. The results are the opposite of those intended: everyone thinks, "if the senior managers are not interested, the problems can't be important."

Whilst managers can make errors in their own tasks their failure to recognise wider safety threats means they do not create the conditions needed within the organisation to manage risks. There appear to be two reasons for this [12]:

1. They are reassured when they see written systems and procedures, even if they don't have any evidence that shows that they are effective in practice. This was the case with the Piper Alpha disaster. The procedures describing the Permit to Work system complied with all the latest good practice but in reality it was not working as intended (see accident summaries).
2. A good safety record gives a false sense of security. This was the case at BP Texas City where a good personal safety record was celebrated whilst process safety was being neglected (see accident summaries).

### 6.2.6.1 Management competence

Kletz said that improving management of risk has to focus on the competence of managers in terms of selection and recruitment, training, and career development. This has to go to the very top of the organisation. Directors and board members need to provide appropriate leadership based on a thorough understanding that health and safety is integral to business success. The most senior personnel must be actively involved and take personal responsibility [13].

One of the challenges for organisations working with major hazards is that directors and board members often have insufficient knowledge of the risks and how they can be managed, including human factors. They tend to respond to events when they happen. This means major accidents and process safety receive less attention because these types of event occur very infrequently.

### 6.2.6.2 Pemex pipeline accident

On 18 January 2019 a leak of gasoline from a pipeline owned by Pemex exploded in the Mexican town of Tlahuelilpan killing at least 131 people and injuring dozens more [14]. The cause of the leak was identified as illegal tapping for theft of fuel. A crowd had congregated to steal fuel, which explained the high death toll. An audit of Pemex's arrangements [15] for monitoring, maintaining, and protecting its pipelines identified a number of significant management failings that indicate there was a general lack of understanding of the risks. They included:

- Failure to install monitoring in all sections of pipework where it had been planned for;
- Failure to repair sabotaged monitoring equipment;
- No procedures in place to maintain and update the SCADA system in a timely manner;
- Inadequate security measures in some locations;
- No formal protocols for responding to and reporting pressure drops on the pipeline, which could be an indication of fuel theft.

Kletz addressed this as follows [2].

### Text Box 6.12

If output, costs, efficiency and product quality fail, the results are soon apparent and obviously require immediate attention. In contrast, if safety standards fall, it may be a long time before a serious accident occurs. The falls in standards is hidden or latent. This may be the reasons why safety is given insufficient detailed attention and instead exhortation to work safely replaces considerations of the real problems. Another reason is that senior managements want a single measure of safety performance. The lost-time accident rate seems to supply this, but does not.

### 6.2.6.3 Modern business

Something that has changed significantly since Kletz's days at ICI is that companies are less likely to have centralised technical and support departments. Sites are expected to be self-sufficient and this can lead managers and technical staff to make decisions based on their own experience and preferences, without referring to others. Whilst Kletz may not have predicted the changes that would take place in business he did recognise the potential issues with localising safety management [2].

### Text Box 6.13

An explosion, which killed four men, occurred in a plant which processed ethylene at high pressure. A leak from a badly-made joint was ignited by an unknown cause. After the explosion many changes were made to the standard of joint-making. The training, tools and inspection were all improved.

Poor joint-making had been tolerated for a long time before the explosion because all sources of ignition had been eliminated and so leaks could not ignite, or so it was believed. The plant was part of a large group but the individual parts were technically independent. The other plants in the group had never believed that leaks of flammable gas will not ignite. They knew from their own experience that sources of ignition are liable to turn up, even though we do everything we can to remove known sources, and therefore strenuous efforts must be made to prevent leaks. Unfortunately the mangers of the ethylene plant had hardly any technical contact with the other plants, though they were not far away; handling flammable gases at high pressure was, they believed, a specialised technology and little could be learnt from those who handled them at low pressure. The factory was a monastery, a group of people isolating themselves from the outside world. The explosion blew down the monastery walls.

If the management of the plant where the explosion occurred had been less insular and more willing to compare experiences with other people in the group, or if the managers of the group had allowed the component parts less autonomy, the explosion might never have occurred. It is doubtful if the senior managers or the group ever realised or accepted or discussed the need for a change in policy. The leak was due to a joint and so joints must be made correctly in future. No expense was spared to achieve this aim but the underlying weaknesses in management system went largely unrecognised. However, some years later, during a recession the various parts of the group were merged.

Similarly, the official report on the King's Cross fire said that there was "little exchange of information or ideas between departments and still less cross-fertilisation with other industries and outside organisations."

### 6.2.6.4 *Mechanisms for sharing good practice*

Although changed business models may have made it more difficult to share information and good practices within an organisation, advances in technology including the internet should provide a useful mechanism for communicating with a wider audience. However, fears about confidentiality often mean people can be reluctant to engage in activities where they may have to share information. This is unfortunate because in most walks of life it is very clear that the more you are prepared to share the more you get back in return. But maybe the largest hurdle is that managers and others do not recognise that they need more information or to learn from others. This is particularly the case when they do not have a technical background or have not been educated in process safety.

### 6.2.6.5 *Chronic unease*

If people recognise a need to learn from others, barriers to communication are likely to be overcome. Observations about management behaviour have led to the suggestion that action can be taken to change perspectives. An example of this is 'Chronic unease', which is a principle that seems to have been effective at communicating to managers (and others) that they need to avoid becoming complacent about the risks of major accidents. Also, that they do not have all the answers and need to involve more people in identifying and solving problems.

People are usually easily satisfied when things appear to be good and this can cause them to relax or focus in other areas. So if they do not receive any clear and convincing information about a risk they are liable to assume it either does not exist or is being managed effectively. Achieving a state of 'chronic unease' encourages people to do more to check what is happening, probing deeper to find out what is really happening.

This can also apply at the operator level. Chronic unease can drive operators to cross check data from different sources to make sure that what they think is happening is actually happening. This can include predicting what they expect to see such as calculating when a vessel should be full, or when a temperature should be reached, so that if it is not trending as anticipated they can respond early whilst the problem is small instead of waiting for an alarm or trip, which usually means the problem is greater.

Chronic unease turns the cliché 'no news is good news' on its head to say that the fact you are receiving no news probably means that the important messages are simply not getting to you. Also, it points out that complex systems are very difficult to understand and so you cannot assume that the information you are receiving is telling you what you think it is. Fundamentally, chronic unease is about changing from the 'normal' satisfaction from receiving good news to a viewpoint that there can always be very big problems but they are often hidden so deeply in the system that you cannot see it unless you look very carefully for them.

#### 6.2.6.6 *Major accidents happen*

A consistent theme from major accidents is that, before they occurred, managers usually did not have any particular concerns about safety. They thought everything was, whilst not perfect, safe enough. However, investigations after the event show that there were usually plenty of problems, but managers had either not looked for them or had not appreciated the potential to cause accidents [12]. Managers should not expect people to tell them about problems. They should assume problems exist and continually be looking for evidence they can use to drive improvement.

## 6.3 Reducing human error probability

The reality is that everyone commits errors and they can never be eliminated completely. Recognising them and understanding their causes gives us the opportunity to reduce the associated risks. This is the focus of human factors (also known as ergonomics), which has emerged as a significant discipline in the years since Kletz started writing about human error.

### 6.3.1 Human strengths and weakness

All humans have inherent strengths and weaknesses. Understanding these allows us to develop systems that minimise the probability of human error as illustrated in Kletz's story as follows [2].

---

### Text Box 6.14

A man went into a tailor's shop for a ready-made suit. He tried on most of the stock without finding one that fitted him. Finally, in exasperation, the tailor said, "I'm sorry, sir. I can't fit you; you're the wrong shape."

Should we as engineers expect people to change their (physical or mental) shapes so that they fit into the plants and procedures we have designed or should we design plants and procedures to fit people?

We know that they cannot change their physical shape. If a man cannot reach a valve we do not tell him to try harder or grow taller. We provide a step, move the valve or remove the need for the valve. But some people expect others to change their mental shape and never have slips or lapses of attention. This is as difficult as growing taller. We should instead change the design or method of working to reduce the opportunities for error.

These principles are known and 'human centred design'. It is a recognition that people make fewer errors when they are given systems that match their physical and mental capabilities.

## 6.3.2 Performance influencing factors

One of the key human factors concepts is the idea of Performance Influencing Factors (PIFs) that affect people and the way they perform at work. At a high level they relate to the Job, Individual, and Organisation [16].

The challenge, especially for engineers who generally prefer objective measures of success, is that there is rarely a correct or incorrect human factors solution. A feature that has a positive effect on one aspect will often increase another risk and so design has to be a compromise. The aim is to optimise the PIFs so that human error probability is reduced whilst recognising that elimination is rarely possible and residual risks have to be understood and managed.

Some of the key methods of improving PIFs to reduce human errors are discussed as follows.

### 6.3.2.1 Written instructions

Most companies will have many written instructions. They are generally written with the aim of ensuring tasks are performed in a safe and effective way. In practice they often fail to achieve this objective. Kletz provided a number of suggestions for how they can be improved [2].

## Text Box 6.15

Here are some questions that should be asked about instructions.

- Are they easy to read?
- Are they written to help the reader or to protect the writer?
- Are they explained to those who will have to carry them out?
- Are they maintained?

**Are they easy to read?**

Many instructions are not. Men are remarkably good at detecting meaning in a smog of verbiage but they should not be expected to do so. Sooner or later they will fail to comprehend.

Are they written to help the reader or to protect the writer? It is not difficult to recognise instructions written to protect the writer. They are usually written in a legalistic language, are long and go into excessive detail. As a result they are often not read. An instruction that covers 99% of the circumstances that might arise, and is read end understood, is better than one that tries to cover everything but is not read or not understood.

**Are they explained to those who will have to carry them out?**

On one works the instructions on the procedure to be followed and precautions to be taken before men were allowed to enter vessel or other confined space ran to 23 pages plus 33 pages of appendices; 56 pages in all. There were many special circumstances but even so this seems rather too long. However, when the instruction was revised it was discussed in draft with groups or supervisors and the changes pointed out. This was time-consuming for both supervisors and manager but was the only way of making sure that the supervisors changes and for the managers to find out if the changes would work in practice. Most people, on receiving a 56-page document, will put it aside to read when they have time - and you know what that means. New instructions should be discussed with those who will have to carry them out.

**Are they maintained?**

Necessary maintenance is of two sorts. First, the instructions must be kept up-to-date and, second, regular checks should be made to see that the instructions in the control room are in a legible condition. If too worn they should obviously be replaced. If spotlessly clean, like poetry books in libraries, they are probably never read and the reasons for this should be sought. Perhaps they are incomprehensible.

A senior manager visiting a control room should ask to see the instructions - operating as well as safety. He may be surprised how often they are out-of-date, or cannot readily be found or are spotlessly clean.

One explosion occurred because an operator followed out-of-date instructions he found in a folder in the control room.

Finally, a quotation from H.J. Sandvig[a]:

> Operators are taught by other operators and ... each time this happens something is left unsaid or untold unless specific operating instructions are provided, specific tasks are identified and written and management reviews these procedures at least annually and incorporated changes and improvements in the process.

Unfortunately problems with written instructions are very prevalent. It is rarely a lack of instructions but that they do not support people when performing tasks and so fail in their main aim of reducing human error probabilities.

Kletz shared the following story [2].

## Text Box 6.16

A plant contained four reactors in parallel. Every 3 or 4 days each reactor had to be taken off line for regeneration of the catalyst. The feed inlet and exit valves were closed, the reactor was swept out with steam and then hot air passed through it. One day a fire occurred in the reactor during regeneration. Afterwards the staff agreed that the steam purging should be carried out for longer and should be followed by tests to make sure that all the feed had been swept out. An instruction was written in the shift handover log and a handwritten note was pinned up in the control room but no change was made to the folder of typed operating instructions.

[a] H.J. Sandvig, Journal of American Oil Chemists Society (JAOCS 60(2)), 1983.

A year later an operator who had been on loan to another unit returned his old job. He saw that the note on extra sweeping out and testing had disappeared and assumed that the instruction had been cancelled. He did the task without any extra sweeping out or any tests. There was another, and larger, fire. There was no system for maintaining instructions. (In addition, the contents of the instructions was inadequate. Feed may have been entering the reactor through a leaking valve. The reactor should have been isolated by slip-plates or double block and bleed valves.)

### 6.3.2.2 Fundamental issues with written instructions

As well as the advice earlier, Kletz alluded to the fact that there are often more fundamental issues with the content of instructions [2].

---

### Text Box 6.17

Other common weaknesses in instructions

- They do not correspond to the way the job is actually done.
- They contain too much or too little detail. Complex unfamiliar tasks with serious results if carried out incorrectly need step-by-step instructions. Simple familiar tasks may be tools of the trade and need no instructions at all. Different levels of detail are needed by novices and experienced people. For example, if a line has to be blown clear with compressed gas, novices should be told how long it will take and how they can tell that the clear.
- The reasons for the instructions are not clear.
- The boundaries of space, time and job beyond which different instructions apply are not clear.

---

The main failing at most companies is that the purpose of instructions is not clear and so there is no way of objectively determining whether they are effective. Also, there is an exaggerated view of what writing a procedure or instruction can achieve.

### 6.3.2.3 Written instructions will always have limitations

The reality is that written instructions appear relatively low on the hierarchy of risk controls (see Chapter 4). It is not possible or desirable to write instructions for every task and a significant downside of writing more is that it becomes more difficult for people to find the ones they need.

The writers of instructions always have to base the content on an assumed set of conditions. In practice the conditions under which tasks are performed will vary and written instructions can rarely cover every eventuality.

Managers will often ask that instructions are written so anyone can follow them—the mythical 'man or woman in the street'. To achieve this they have to cover every detail. When competent people come to use the instructions they find that there is far too much trivial information and quickly recognise it is not meant for them. Also, this extra detail significantly increases the probability that critical information is overlooked. As a rule of thumb, writing an instruction so that someone competent in the role (e.g. operations, maintenance) but from another site can understand how to perform the task can guide what to include and what to leave out.

### 6.3.2.4 A general philosophy

As a general rule of thumb the following can be used to determine how instructions should be developed:

- High criticality, complex tasks performed infrequently—Mandatory 'Print-Follow-Sign' instructions providing a full step by step task instruction that has to be actively used each time the task is performed;
- Medium criticality, moderately complex tasks performed relatively frequently—Reference instructions providing a detailed description of the task that is available to support training and as a refresher, but people deemed competent can perform the task without reading the instruction each time provided they follow the prescribed method;
- Low criticality, simple tasks performed very frequently—No written detailed instruction provided. Less detailed or more general guidance may be provided, and people will learn on-the-job and demonstrate their competence through practical demonstration.

Effort should be focused on the high criticality tasks, which should typically make up less than 20% of the total. A systematic process should be followed to identify them (for example the method described in the HSE report 1999/092 [17]).

### 6.3.2.5 Instructions for the wrong tasks

It is very common to find that organisations with many written instructions do not have any to cover the most critical tasks. One of the reasons for this is that these tasks are performed relatively infrequently and so the need is not always recognised. Of course it is the tasks performed less frequently where a good written instruction can have most value.

An incident occurred on at a tanker loading facility. It occurred when product was actually pumped off of the tanker (i.e. the opposite of the normal operation) because there was a technical problem with the vehicle. The investigation found that there was a good quality written instruction for loading tankers, which was relatively straightforward because a dedicated loading arm was provided. Although unplanned the requirement to remove product from a tanker was known. However, because it occurred very infrequently it had never been properly considered. This meant the site operators had to develop their own method and improvise using a hose intended for another purpose, which they attached to a drain valve.

In this case effort had been put into writing a written instruction that provided relatively little benefit because the loading arm simplified the task and personnel had plenty of opportunity to learn the method because it was performed so frequently. But a written instruction that would have been of most benefit and may have prevented the incident had not been considered.

### 6.3.2.6 Formatting written instructions

A 'one size fits all' approach to writing instructions is a common problem. Whilst it may suit the writer or quality assurance system it misses the point that the most critical factor is that instructions are only effective if they support the user. Choosing the best format for the instruction and the level of detail should always be based on the nature of the task (e.g. criticality and complexity), the requirements of the end user (typically related to their experience of the task and knowledge of the system), and the circumstances in which the task will be performed (e.g. normal operation, emergency).

Kletz pointed out that instructions do not always have to be simple lines of text [2] whilst emphasising the need for them to be appropriate for the task.

---

### Text Box 6.18

Would a check-list reduce the chance that someone will forget to open a valve? Check-lists are useful when performing an unfamiliar task—for example, a plant start-up or shutdown which occurs only once per year. It is unrealistic to expect people to use them when carrying out a task which is carried out every day or every few days. The operator knows exactly what to do and sees no need for a check-list. If the manager insists that one is used, and they each step is ticked as it is completed, the list will be completed at the end of the shift.

When chokes on cars were manual, we all forgot occasionally to push them in when the engines got hot but we would not have agreed to complete a check-list every time we started our cars.

---

In practice there are a number of formats that can be used depending on the task.

Checklists are particularly useful for people to keep track of where they are up to in the task, which can be particularly critical at shift handover or if there are other pauses in the task or changes of personnel.

Flow charts are useful where decisions have to be made. Diagrams and photos can be useful where it is difficult to describe an activity using text or where users of the instruction will benefit from being able to see what success looks like.

Although these different presentation methods can be very useful they also require more effort to generate than simple text. This emphasises why the format should be appropriate to the task and the user and it is not the case that, for example, a flow chart is always better than a checklist.

#### 6.3.2.7 Emergency instructions

Instructions for handling emergency situations require a different approach. To be effective they must support the people who have to detect, diagnose, and respond to hazardous situations meaning they must be relevant to the hazardous situations that can occur and provide strategies that can realistically be implemented. Also, they must be presented in a way that ensures critical and useful information is communicated to people who will be dealing with a demanding and stressful situation [18]. It must be recognised that the people who usually deal with the initial stages of an emergency are the process operators, particularly in the control room. Emergency response personnel may perform the bulk of the activities once they arrive, but for that to happen they have to be notified by the operators and take time to mobilise.

A false fire alarm in a warehouse due to damaged cables triggered the automatic foam system, engulfing a section of the warehouse. A series of misunderstandings in the emergency response ultimately culminated in a fatality [19] when firefighters entered the area where the foam had been deployed. The foam was viscous, making it difficult for them to move. Also, it restricted vision and acted as sound insulation. Although well trained in using breathing apparatus it was found that the firefighters had used twice as much air when working in the foam than would normally have been expected. In this case the need to enter the area filled with foam was based on a misunderstanding of risks and the implications of working in an area filled with foam were not recognised. The emergency procedures being followed did not support the people to make appropriate and safe decisions.

### 6.3.2.8 *Content is king*

Whilst being clearer about the role of written instructions and making sure the best format is used the most important factor is that they describe the correct method of performing the task. This means that the task will achieve its objectives in a safe way, but also it must be practical to perform. One of the most common complaints of users is that written instructions do not reflect how tasks are performed in practice.

To get the content correct it is essential to actively involve the people who perform the task in the initial writing, to walk through the task to validate the written instructions before use, and to receive feedback when the task is performed.

Avoiding overly prescriptive written instructions helps to ensure that they can be used in all appropriate circumstances. Also, this can be effective at minimising the text which helps readability and ensures focus on the most important information.

Whilst instructions can make a valuable contribution to management of risks it must be recognised that they appear low on the hierarchy of risk controls (see Chapter 4). Instructions should only be put in place to manage the residual risks after more reliable means including hazard elimination and reduction, and engineering controls have been implemented.

## 6.3.3 Training and competence

Written instructions clearly only have value if they are used as intended. Kletz highlighted that training was an important aspect in this [2].

---

### Text Box 6.19

Training gives us an understanding of our tasks and equips us to use our discretion, while instructions tell us precisely what we should and should not do; training equips us for knowledge-based and skill-based behaviour while instructions equip us for rule-based behaviour. Which do we want?

---

Written instructions should be developed to support competent people and this requires an understanding of the levels of competence to ensure instructions are appropriate for the user.

### 6.3.3.1 *Training vs competence*

Training is only one factor in people becoming competent.

Competence can be defined as the ability to do something successfully. For someone to be considered competent they need to have the required skills, knowledge, understanding, and attitude.

Competence is not a single achievement that lasts for ever. People who may be considered competent enough to do a job can become more competent over time by gaining practical experience. This can allow them to work more efficiently and effectively, and be better able to handle more complex situations that arise. Equally, people who may have been deemed competent at one time may lose this due to insufficient opportunities to maintain experience on particular aspects of the job, due to becoming complacent or failing to appreciate that changes made to the system require them to change the way the job is performed.

### 6.3.3.2 Competence management systems

Competence is probably one of the most important factors in reducing human error probabilities but it is often addressed in an unsystematic way. This can be overcome by developing a competence management system that identifies the safety critical roles within the organisation and the competencies required for each.

Achieving competence requires a mixture of qualifications, vocational training, experience, and self-development. It should start when selecting people for roles and direct initial and follow-up on-the-job training. Once someone is considered competent for a role the way that they maintain and develop it further needs to be managed.

### 6.3.3.3 Competence assessment

Competence assessment has to be an integral part of the management system and should focus on the competencies required for the job rather than simply confirming that someone has been trained. However, it can be an emotive subject with people fearful that their job could be in jeopardy if they 'fail' an assessment. It needs to be handled carefully and the people involved should understand that a lack of competence is an indication of a failed system and not of them personally.

Competence assessment should not be viewed as a means of deciding if someone is capable of performing their job. It should recognise that there are usually different levels of competence and to become fully competent usually requires experience and exposure to a range of tasks and situations that are often outside the control of the individual being assessed. If assessment highlights gaps in competence the main questions should be why has the management system failed to support the individual and what can be done differently to allow them to fill those gaps?

### 6.3.3.4 Training courses do not make people competent

It is easy to assume that someone attending a training course that includes an assessment and results in a certificate being issued is enough for someone to become competent. Whilst it may be effective at providing a step change in someone's competence it is rarely enough for them to be fully competent in performing that role at the workplace.

As an example, anyone installing or maintaining electrical equipment in an area where there is a potential risk of explosive atmospheres must be suitably competent. There are recognised certification schemes that provide training and assessment based on an approved curriculum. But the focus of these courses is on the avoidance of ignition and it cannot demonstrate that people have all the competencies they need to be an electrician at a facility that handles flammable materials.

Before attending the course the individual should already be a competent electrician. Without this they may know about the safety aspects taught on the course but will still not be able to do the job.

Following the course the person needs to have the opportunity to put their newly acquired skills into practice in the workplace. This should be under close supervision until it is confirmed that they are able to apply what they have learnt.

Having been deemed fully competent the person should continue to practice the skills otherwise their competence will degrade. It is never the case that anyone can be considered competent forever and this should be reflected in every competence management system.

Having a certificate does not, on its own, prove competence.

### 6.3.4 Communication

Communication is a key source of information that people need to do their job. The nature of language means that human error is a natural part of communication and so particular care is required when safety critical information is being communicated.

Although communication takes place continually certain activities have been identified as particularly critical: shift handover, permit to work, and emergency response.

#### 6.3.4.1 *Communication at Piper Alpha*

Poor communication at shift handover and a breakdown of the Permit to Work system were identified amongst the causes of the Piper Alpha disaster (see accident summaries). The operators decided to start the standby condensate pump but did not realise its relief valve was not in place. The inquiry into the disaster found no evidence to suggest that the people involved had done this intentionally and concluded that they made decisions that, in hindsight, were clearly wrong because they did not have a full and accurate understanding of equipment status and condition [20]. People need to be guided about what information to communicate and encouraged to prepare in advance, take the time required, and ensure all parties are actively involved. Although lack of information is usually the main concern, attempting to communicate too much trivial detail can hinder the overall communication process.

#### 6.3.4.2 *Behavioural aspects of communication*

There is a very significant behavioural aspect to communication. To work effectively the people involved need to be willing to say if they do not understand what they have been told and challenge what they have been told. No procedure or management system can address these issues directly. Whilst guidance can be provided to help people understand what is expected of them, there will be a requirement for continuous supervision and coaching to ensure bad habits are avoided and to drive continual improvement.

## 6.3.5 Control room design

Advances in technology have resulted in the control room operator role becoming increasingly critical to safety. Control room operators need to have access to the information they need to keep systems safe and they need to be alert and motivated to detect potential problems early and respond appropriately. Control room is design concerned with the Human Machine Interfaces (HMI), the working environment, and other arrangements that affect communication and human performance.

### 6.3.5.1 *Human machine interfaces*

Kletz's following example [2] illustrates how operators can be overloaded and why it is important that this is avoided through design of the HMI.

---

**Text Box 6.20**

Information or task overload.

A new, highly automated plant developed an unforeseen fault. The computer started to print out a long list of alarms. The operator did not know what had occurred and took no action. Ultimately an explosion occurred.

Afterwards the designers agreed that the situation should not have occurred and that it was difficult or impossible for the operator to diagnose the fault, but they then said to him, "Why didn't you assume the worst and trip the plant? Why didn't you say to yourself, 'I don't know what's happening so I will assume it is a condition that justifies an emergency shut-down. It can't be worse than that'?"

Unfortunately people do not think like that. If someone is overloaded by too much information he may switch off (himself, not the equipment) and do nothing. The action suggested by the designers may be logical, but this is not how people behave under pressure.

The introduction of computers has made it much easier than in the past to overload people with too much information, in management as well as operating jobs. If quantities of computer print-out are dumped on people's desks every week, then most of it will be ignored, including the bits that should be looked at.

---

People are able to handle a lot of information provided it is presented to them in an appropriate way. Unfortunately too many HMI displays present data in a numerical form, often included as part of simple schematic diagrams that have been copied directly from piping and instrument diagrams. This means that the displays are not focussed on providing the operator with the information they need to monitor plant or perform tasks reliably and effectively. Also, this method of displaying information is not consistent with human capabilities, which include the ability to recognise shapes and patterns.

Trend displays are particularly effective at allowing control room operators to monitor processes, allowing them to identify subtle changes over time. Other display types such as virtual gauges and radar plots can be used to good effect [21]. Colour is another consideration and its use should convey meaning in a clear and unambiguous way.

### 6.3.5.2 *Alarms*

Another failing of many HMI is that alarms cause nuisance to the control room operator and sometimes overload them with data. Kletz highlighted this problem [2].

---

### Text Box 6.21

Increasing the number of alarms does not increase reliability proportionately.

Suppose an operator ignores an alarm on 1 in 100 of the occasions on which it sounds. Installing another alarm (at a slightly different setting or on a differ parameter) will not reduce the failure rate to 1 in 10,000. If the operator is in a state in which he ignores the first alarm, then there is a more than average chance that he will ignore the second. (In one plant there were five alarms in series. The designer assumed that the operator would ignore each alarm on one occasion in 10, the whole lot on one occasion in 100,000!).

---

It has become very easy to add alarms to the control system and this has meant too many have been configured on most systems. Before any alarms are considered a clear philosophy should be developed that explains that alarms should only be used to warn of potentially critical situations requiring timely intervention by the control room operator. Any other events and indications, including where the operator cannot respond in time or the event has already happened (including process trips) should not be an alarm. Alarms should be prioritised according to the potential consequences of no response and the time available to respond [3].

### 6.3.5.3 *Alertness and motivation*

Kletz identifies another potential problem in control rooms, that being task underload [2].

---

### Text Box 6.22

Reliability falls off when people have too little to do as well as when they have too much to do. It is difficult for night-watchmen to remain alert.

During the Second World War, studies were made of the performance of watch-keepers detecting submarines approaching ships. It was found that the effectiveness of a man carrying out such a passive task fell off very rapidly after about 30 min.

It is sometimes suggested that we should restrict the amount of automation on a plant in order to give the operators enough to do to keep them alert. I do not think this is the right philosophy. If automation is needed to give the necessary reliability, then we should not sacrifice reliability in order to find work for the operators. We should look for other ways of keeping them alert. Similarly if automation is chosen because it is more efficient or effective, we should not sacrifice efficiency or effectiveness in order to find work for the operators.

In practice, I doubt if process plant operators often suffer from task underload to an extent that affects their performance. Although in theory they have little to do on a highly automated plant, in practice there are often some instruments on manual control, there are non-automated tasks to be done, such as changing over pumps and tank, there is equipment to be prepared for maintenance, routine inspections to be carried out, and so on.

---

If, however, it is felt that the operators are seriously under-loaded, then we should not ask them to do what a machine can do better but look for useful but not essential tasks that will keep them alert and which can be set aside if there trouble on the plant - the process equivalent of leaving the ironing for the babysitter.

One such task is the calculation and graphing of process parameters such as efficiency, fuel consumption, catalyst life and so on.

---

As Kletz suggests it may not be caused by underload, but operators often struggle with alertness and their physical and mental wellbeing can be at risk. This is a concern because the critical nature of the role means that we need our control room operators to be fully attentive, focussed, and motivated at all times.

Unfortunately, the working environment created by many control rooms achieves the opposite. Some are unpleasant places to work because of their physical design. In other cases the air quality or lighting is not fit for purpose. Control room operators are generally very good at adapting to their environment and most of the time the system operates as intended and so their actions are not of immediately critical importance. However, if things start to go wrong any delay in detecting, diagnosing, or responding to the problem can allow escalation into something far more serious.

Control room design should be viewed as a safety critical requirement [22]. It is not good enough that control room operators are able to cope during normal operations because it is their ability to handle unplanned, complex, and challenging events that has the biggest impact on process risk.

## 6.3.6 Fatigue management

Kletz identified the affect that fatigue can have on human error probability [2].

---

**Text Box 6.23**

Fatigue may make slips, and perhaps errors of other sorts, more likely. In the Clapham Junction railway accident a contributory factor to a slip was 'the blunting of the sharp edge of close attention' due to working 7 days per week without a day off. The man involved was not, however, physically tired.

The report on a chemical plant accident (according to press reports a flare stack was filled with oil which overflowed, producing flames several hundred feet high) said, "It is clear that in this case operating teams overlooked things that should not have been overlooked, and misinterpreted instrument readings. A major influence over the behaviour of the operating teams was their tiredness and frustration." A trade union leader is quoted as saying that the management team members were more tired than the operators as the managers were working 12-h shifts.

Obviously we should design work schedules that do not produce excessive fatigue, but inevitably some people will be tired from time to time as the result of factors that have nothing to do with work. When we can we should design plants and methods of working so that fatigue (like other causes of error) does not have serious effects on safety, output and efficiency.

---

The influence of fatigue on human performance and means of controlling the associated risk has been subject to many studies. Shift work creates particular concern but it is not the only cause of fatigue, and working patterns of everyone in a potentially safety critical role need to be carefully controlled.

### 6.3.6.1 Consequences of fatigue

The two main concerns of fatigue are as follows:

- Reduced alertness affecting the ability to detect situations, make decisions, and act reliably;
- Chronic effects that can result in ill health to the individual and affect their physical and mental capability to do their job.

It has been suggested that performance after being awake for 18 hours is equivalent to a drunk driver being over the drink driving limit.

Methods of evaluating work and shift patterns have been developed including the 'Fatigue Index' published by the UK HSE [23]. However, this is often a too simplistic evaluation because the human factors risks go beyond fatigue and include issues such as communication between teams (e.g. shift handover), continuity of operation and control, and ability to cover short-term absence (e.g. sickness).

### 6.3.6.2 Managing fatigue

Fatigue cannot be eliminated, especially when people work in shifts. Although the hours spent at work is a factor, it is the ability to recover effectively between shifts that is often more significant.

Guidance [24] suggests that shift patterns should restrict the number of night shifts to less than four wherever possible and that the rotation should always be 'forward' so that people work days followed by nights; or mornings, then afternoons then nights.

Shift workers should be encouraged to follow a healthy lifestyle inside and outside of work and be advised about developing good sleeping habits.

Breaks during the working day are important for restoring alertness and should be properly managed and not left to the individual to organise themselves. Being allowed to take short naps has been identified as a very effective way of overcoming short-term fatigue issues.

Arranging for more interesting and varied work to be done at night and at other low points should be a management responsibility.

Day workers can also suffer from fatigue. Even if their normal workload is kept manageable so that they do not have to work overtime routinely, periods of high workload due to projects or breakdowns can result in many extra hours being worked. This can be particularly problematic if this is additional work, meaning that they still have to find time to carry out their 'normal' tasks. Also, it must be recognised that day workers' breaks are generally fixed (typically weekends) and if this is lost due to extra work there can be little opportunity for any compensatory rest.

## 6.3.7 Staffing arrangements

One topic of interest or even fascination for many years has been how can you identify the correct number of people in an organisation or ensure an appropriate level of workload. Kletz highlighted why insufficient people or excessive workload for people at work results in risk [2].

## Text Box 6.24

Plant supervisors sometimes suffer from task overload—that is, they are expected to handle more jobs at once than a person can reasonably cope with. This has caused several accidents. For example, two jobs had to be carried out simultaneously in the same pipe trench, 20 m apart. The first job was construction of a new pipeline. A permit-to-work, including a welding permit, was issued at 08.00 h, valid for the whole day. At 12.00 h a Permit was requested for removal of a slip-plate from an oil line. The foreman gave permission, judging that the welders would by this time be more than 15 m (50 ft) from the site of the slip-plate. He did not visit the pipe trench which was 500 m away, as he was dealing with problems on the operating plant. Had he visited the trench he might have noticed that it was flooded. Although the pipeline had been emptied, a few gallons of light oil 'thawed and ran out when the slip-plate joint was broken. It spread over the surface of the water in the pipe trench and was ignited by the welders. The man removing the slip-plate was killed.

The actual distance between the two jobs—20 m—was rather close to the minimum distance—15 m—normally required. However the 15 m includes a safety margin. Vapour from a small spillage will not normally spread anything like this distance. On the surface of water, however, liquid will spread hundreds of metres.

Afterwards a special day supervisor was appointed to supervise the construction operations. It was realised that it was unrealistic to expect the foreman, with his primary commitment to the operating plant, to give the construction work the attention it required.

Despite claims by some, there is no simple formula that allows you to calculate the correct number of people. Benchmarking has been attempted, but this requires having an appropriate comparisons. The reality is that there are far too many variables to allow this.

Although difficult, it does not mean that organisations do not have to attempt to evaluate their staffing levels and arrangements. What is required is an objective view of how well individuals and teams will be able to cope in the highest demand situations and what would be the consequences if they did not cope. For hazardous operations the most interesting scenarios are usually the early stages of an event that can escalate if it is not properly and promptly detected, diagnosed, and responded to. This is where human intervention can make the most difference. If escalation does occur the human actions are generally restricted to evacuation and recovery.

The reason why more 'normal' activities are generally less of interest than the high demand situations is that insufficient staffing levels will generally mean that activities are simply not performed or delayed and immediate consequences are minimal. However, this can be a significant performance influencing factor and so care is required to ensure that staffing pressures do not result in short cuts being taken or elective tasks such as preventative maintenance being overlooked.

The UK National Health Service is often cited as an example of an organisation that is continually lacking in staff. When people present with immediately life-threatening conditions the system is able to respond but people arriving at accident and emergency departments with less serious conditions have to wait to be treated, as there are long waiting lists for operations and frequent cancellations. This may be due to poor management or underfunding from central government, but we see similar situations in industry where people are able to respond to breakdowns and production problems but struggle to find (or make) time to do the more routine tasks intended to avoid these problems in the first place.

## 6.3.8 Signs and labels

Kletz identified the importance of labelling [10].

---

### Text Box 6.25

Many incidents have occurred because equipment was not clearly labelled. Some of these incidents have already been described in the section on the identification of equipment under maintenance. Seeing that equipment is clearly and adequately labelled and checking from time to time to make sure that the labels are still there is a dull job, providing no opportunity to exercise our technical or intellectual skills. Nevertheless, it is as important as more demanding tasks are. One of the signs of good managers, foremen, operators, and designers is that they see to the dull jobs as well as those that are interesting. If you want to judge a team, look at its labels as well as the technical problems it has solved.

---

### 6.3.8.1 *Problems caused by poor labelling*

Kletz identified a number of instances where poor labelling has caused problems including [10]:

- Indication of damper status, whether it was open or closed;
- Incorrect labels on electrical fuses or switchgear resulting in the wrong equipment being isolated before work commenced;
- Sample points not identified and so the wrong sampling and testing procedures are followed;
- Connecting to a low pressure service line allowing back-flow of process into service supply system;
- Filling tankers with the wrong product;
- Lifting devices labelled with the wrong capacity;
- Confusion about units being shown on gauges;
- Confusion about chemical identity.

### 6.3.8.2 *Loading oxygen instead of nitrogen*

A potentially very serious incident occurred when filling tankers with nitrogen [10].

---

### Text Box 6.26

A tank car was fitted with nitrogen connections and labelled Nitrogen. Probably because of vibration, one of the hinged boards fell down so that it read Oxygen. The filling station staff therefore changed the connections and put oxygen in the tank car. Later, some nitrogen tank trucks were filled from the tank car—which was labelled Nitrogen on the other side—and supplied to a customer who wanted nitrogen. The customer off-loaded the oxygen into his plant, thinking it was nitrogen. The mistake was found when the customer looked at his weigh-bridge figures and noticed that on arrival the tanker had weighed 3 tons more than usual. A check then showed that the plant nitrogen system contained 30% oxygen.

### 6.3.8.3 *Signs vs labels*

Signs and labels perform different purposes. A sign should convey information and a label is primarily for identification. For example, it would be good practice for every valve on plant to have a unique identification number which will be shown in the field via a tag or similar. This will allow it to be positively identified by an operator or maintenance technician, allowing them to cross check with P&ID, instructions, etc. However, if identification of that valve is particularly critical or its status has to be controlled in a particular way it may be appropriate to provide a sign in addition to the identification label. This would be displayed prominently and its design should ensure it is legible from the position where the valve will be operated.

### 6.3.8.4 *Maintaining signs and labels*

One of the common problems with signs and labels is that they become damaged, weathered, painted over, or are removed during maintenance and are not replaced. They should be subject to routine inspection and maintenance to ensure these problems are rectified.

It is generally accepted at original design and construction that signs and labels are critical. However, few organisations have an effective mechanism for formally recognising this and incorporating the requirement into their maintenance management system. The result is that many sites have allowed signs and labels to disappear over the years, which can have a significant impact on human error probability.

## 6.3.9 Behavioural safety

Most of this chapter has been focussed on reducing the likelihood of error through better design and systemic changes. This can cause people to worry that it takes away personal responsibility. This is something Kletz was aware of [2].

### Text Box 6.27

The reader who has got this far may wonder what has happened to the old-fashioned virtue of personal responsibility. Has that no part to play in safety? Should people not accept some responsibility for their own safety?

We live in a world in which people are less and less willing to accept responsibility for their actions. If a man commits a crime it is not his fault, but the fault of those who brought him up, or those who put him in a position in which he felt compelled to commit the crime. He should not he blamed, but offered sympathy. If someone is reluctant to work, he or she is no longer work-shy or lazy but a sufferer from chronic fatigue or some other recently discovered syndrome.

This attitude is parodied in the story of the modern Samaritan who found a many lying injured by the side of the roadside and said, "whoever did this to you must be in need of help." And in the story of the schoolboy in trouble who father, "What's to blame, my environment or my heredity?" Either way it was not him.

Many people react to this attitude by re-asserting that people do have free will and are responsible for their actions. A criminal may say that his crimes were the result of present or past deprivation, but most deprived people do not turn to crime.

---

One approach to reducing accidents caused by human error has been Behavioural Safety. This is usually a programme that aims to change the way people think about their work and identifying unsafe behaviour that people need to avoid. It usually involves observation of people working where the observer makes note of any unsafe acts and gives feedback after the work is complete.

There can be many positive outcomes from implementing a behavioural safety programme, and there have been many claims that accident rates have fallen quite dramatically as a result. However, they require a lot of resource and there is an argument to say that if this was directed to improving design, management, and systems the results would be far more wide ranging and sustainable, especially for process safety. Behavioural safety is certainly a potentially useful addition to the safety toolkit [25] but it should not be considered as an alternative to the other, systems-based approaches to human factors discussed in this chapter.

## 6.4 Human factors analysis

Kletz was instrumental at highlighting the importance of human error in safety and that human factors have to be an integral part of all risk analyses.

### 6.4.1 Quantitative analysis

Kletz was quite keen on the idea of quantifying the probability of human error. His view was that humans are part of the total protective system. If we want to know the reliability of the system we need to know the reliability of each component including the human.

Many attempts have been made over recent decades to develop methods for quantifying human error probability but it has proven to be problematic because humans are subject to so many influencing factors that their reliability can vary dramatically from one situation to another. This can mean the reliability of the same person performing the same task can vary significantly from 1 day to another (or even hour to hour).

#### 6.4.1.1 A conservative approach

Where quantifying human error probability is likely to provide some benefit the preferred approach is to take very conservative estimates. The main purpose is to allow sensitivity analysis to take place, so that the effect of human reliability on the whole system performance can be determined (see Chapter 3).

A probability of human failure of 0.1 per operation is widely used in Layers of Protection Analysis (LOPA) as a default value given this is referenced by Center for Chemical Process Safety [26]. This should not be viewed as the actual likelihood of someone making an error when performing a task but can be used to explore the sensitivity of the overall assessment to the direct and indirect human contributions.

The UK's Health and Safety Executive encourages qualitative analysis of human reliability in most instances (see Section 6.4.2). It does allow for quantification as part of a demonstration of risks, but only if performed within strict criteria. It states that "Use of generic Human Error Probability (HEP) data is unacceptable unless it has been qualified to reflect the local circumstances or is more than or equal to an HEP of 0.1" [27]. Again this is not intended as the expected reliability but is intended to be used to determine the relative contribution of human error to overall risk.

Anyone wishing to quantity human error probability, especially if they wish to claim values better than 0.1, needs to be able to demonstrate a high level of competence in human factors and apply sound methods to ensure the values used are reasonable. An objective assessment of the performance influencing factors (PIF) that affect human reliability and the quality of the factors likely to be present at the time someone is performing the task are required.

One of the problems is that not all PIFs are identified and evaluated. For example, during a LOPA, it is common to consider competence and stress as the dominant PIFs. It is then assumed that the person performing the task is competent and unlikely to be stressed at the time. However, it would be very difficult to prove that this will always be the case. Also, this does not take account of all existing (latent) or dynamic (changing) PIFs. For example, for a valve mal-operation, this would not take into account the likelihood of error increasing where, for example, similar valves are arranged in close proximity or out of sequence, and where pump layouts are mirrored and therefore the valves are in different orientation for each pump skid (design-induced errors).

Going into the field to assess how work is actually done with the person who actually performs the task is essential to bring reality to any quantified analysis and reflect the conditions that operators experience by identifying PIFs that make mistakes more likely. It still requires imagination to consider what the PIFs may be like at the time someone is actually performing the task. For example, walking through a task on a summer's day is of limited value if the task may be performed in winter in the middle of the night.

Indicative values for risk factors are available, for example for unfamiliarity, time shortages, non-intuitive design, and information overload [28]. These values show how PIFs influence the potential for mistakes. Understanding the impact of these values can help the front line operations understand the relative magnitude increase in risk associated with conditions in the work place; for example, when an operator is undertaking a task that is unfamiliar, or they are short of time, or the design layout is not intuitive or they are having to take in too much information as part of decision making associated with this or other tasks. This can then influence decisions taken by the front line, especially when more than one of these factors occurs at the same time which can increase risk significantly—in fact, the cumulative risk can be as significant as over-riding a safety instrumented system. What this means for the front line is that where the potential consequence of any failure is high, additional risk assessment should be performed just as would be performed for over-riding a safety instrumented system. This risk assessment would identify additional reduction measures to prevent the potential for mistakes, for example, an additional check by an independent operator may be effective for a complex, critical task that is undertaken infrequently; although achieving full independence of checks like this can be very difficult in practice. In this way the indicative values can be effective at highlighting how risks for front line personnel can be reduced but the calculated risk should not be considered as being accurate unless they have been properly validated for the actual working conditions, which in most cases is likely to be very difficult.

Effective and useful quantified assessment requires a high quality qualitative analysis as its basis. Safety Critical Task Analysis (SCTA) is generally considered to be the most appropriate (see later) provided it includes an evaluation of the overarching approach to risk management to confirm an appropriate application of inherent safety and hierarchy of risk controls (see Chapter 4) and does not simply rely on use of procedures, training, and individual behaviours to manage the risks. To do this requires a detailed analysis of potential errors and PIFs. SCTA is sometimes considered to be a 'pre-accident' investigation – fixing the conditions that can cause an incident before the incident actually occurs (although this could be said about most hazard identification and risk assessment methods).

### 6.4.1.2 *Optimistic claims*

It is sometimes permissible to consider a less conservative value for human reliability provided it can be demonstrated that arrangements are favourable to the person. This must be fully justified by some form of task analysis and an objective analysis of possible performance influencing factors taking into account all modes of operations and scenarios. This is a specialist assessment. However, the most reasonable conclusion may be that there is an overreliance on human reliability and so an alternative solution should be found.

## 6.4.2 Qualitative analysis

The main requirement is to ensure that human factors risks are As Low As Reasonably Practicable (ALARP). In most cases a purely qualitative analysis is sufficient to identify the human errors of main concern and to determine whether existing risk controls are sufficient or more needs to be done.

### 6.4.2.1 *Safety critical task analysis*

Safety Critical Task Analysis (SCTA) has become a standard approach for organisations working in major hazard industries to systematically and objectively consider the potential for human error. It involves identifying the most critical tasks that involve interaction with hazardous system and, due to their nature, are likely to be vulnerable to human error.

Tasks considered to be most critical are subject to a systematic analysis of how they are performed, potential human error, and their consequences. PIFs are evaluated to determine how they will affect the probability of human errors. Two key requirements are to actively involve the people who perform the task when carrying out the analysis and observing or walking through the task to make sure the analysis reflects reality.

When analysing tasks it is important to examine any existing procedures because experience shows that they often do not reflect the way a job is actually performed. Also, to observe the task or at least walk it through where it is performed to determine if there are any actions performed that are physically or mentally demanding. Also, viewing the workplace is an opportunity to detect any evidence of workarounds and short cuts such as indentations in pipework lagging, indicating people may be using it as a step to reach an inaccessible valve, or poles propped in strategic locations showing that they may be used as levers to operate a stiff valve or to defeat a spring-loaded valve.

The Energy Institute has published guidance on performing SCTA [29], including a number of methods that can be used. It references methods intended to be used by task practitioners who are unlikely to have technical expertise in human factors but are better able to integrate the analyses into existing activities on site. For example, a simplified SCTA called the Task Improvement Process [30] provides a systematic way to identify conditions in the work place which make a mistake more likely, so action can be taken to address these PIF, using solutions at the higher end of the hierarchy of controls (preventative, passive, or active controls rather than procedural) where possible. In doing this, potential incidents can be removed from the work place.

Examples of PIFs that make mistake more likely include:

- Conditions which make the activity difficult to undertake, like cramped conditions or lack of visibility (e.g. cannot see level indicator when operating a drain valve or unable to operate a spring-loaded drain valve without crouching for a long period)
- Unclear equipment arrangement and signage (e.g. error traps like two valve arrangements located near each other which look similar or parallel pumps arranged as mirror images; errors in design like these can lead to opening the wrong valve)
- Mundane tasks or tasks of long duration which provide more potential for interruptions and distraction, or where someone could be inclined to do another task at the same time (e.g. where it takes a long time to drain a tank via a manual valve)
- Unfamiliarity with the task, or tasks undertaken infrequently
- Multi-step tasks with potential for doing steps out of order, or missing a step
- Where it is difficult to observe changes in conditions
- Right tools unavailable for a task
- Difficulties in communications between team members.

### 6.4.2.2 Linking with other safety studies

Human factors analyses should link with other safety studies to ensure a consistent view of risk is developed. In particular, tasks identified as either threats or risk control measures in HAZOP, PHA, or and Bowtie diagrams need to be considered. Also, maintenance of engineered controls. Equally, the findings from human factors analyses should be fed back into the other safety studies.

### 6.4.2.3 Safety critical activities

Safety Critical Task Analysis is only applicable for clearly defined tasks. Human interactions that are not executed as tasks may be considered as Safety Critical Activities [31]. Task analysis is not appropriate for these but given their criticality it is important that an appropriate method is used for analysis in order to ensure the associated risks are ALARP. Examples include:

- Routine monitoring of system performance—human reliability will depend on the quality of the HMI and so evaluation of control room design [22] and alarm performance [3] is required;
- General maintenance involving breaking containment—risks are typically handled by the permit-to-work system and so an evaluation of its effectiveness based on available guidance for permit to work [32] and process isolations [33] should be carried out;
- Preventative maintenance—one of the main concerns is that the inspections, testing, and maintenance are carried at the required frequency, which is addressed by other reliability methods. The critical factor is that maintenance schedules are achieved. Some maintenance will be considered as a safety critical task and should be subject to the analysis described earlier;
- Response to incidents and emergencies—systems and procedures need to be in place to support people who may have to deal with these situations. A key element of this is the staffing arrangements [34].

### 6.4.2.4 *Human factors competence*

Organisations handling major hazards should have human factors competence. It may still be viewed as a specialist subject and so this competence may be purchased from consultants, but being an intelligent customer requires some degree of understanding. This is important because management of human factors requires far more than carrying out discrete analyses. It is something that needs attention on a day-to-day basis at all levels of the organisation.

## References

[1] T. Kletz, By Accident—A Life Preventing Them in Industry, PFW Publications, 2000.
[2] T. Kletz, An Engineer's view of Human Error, third ed. published, IChemE, 2001 First published 1985.
[3] Engineering Equipment and Materials Users Association, EEMUA Publication 191 Alarm systems—A Guide to Design, Management and Procurement, third ed., (2013).
[4] A. Brazier, Interlocking Isolation Valves - Less Is More, Hazards 27, Institution of Chemical Engineers, 2017.
[5] U.S. Chemical Safety and Hazard Investigation Board, Vinyl Chloride Monomer Explosion, Formosa Plastics Corp, CSB Investigation Report, (2007).
[6] Anon, Pneumatic Explosion During Work on a Pipe at a Pyrotechnic Plant, Institute of Chemical Engineers, 2016 Loss Prevention Bulletin 248.
[7] U.S. Chemical Safety and Hazard Investigation Board, Ink Dust Explosion and Flash Fires in East Rutherford, CSB Case Study, New Jersey, 2012.
[8] International Association of Oil & Gas Producers, OGP Report 454—Human Factors Engineering in Projects, (2011).
[9] A. Brazier, Human Factors Engineering at the Early Phases of a Project, Contemporary Ergonomics & Human Factors, 2017.
[10] T. Kletz, What Went Wrong?—Case Histories of Process Plant Disasters and How They Could Have Been Avoided, fifth ed., Elsevier, 2009.
[11] U.S. Chemical Safety and Hazard Investigation Board, Key Lessons for Preventing Hydraulic Shock in Industrial Refrigeration Systems Anhydrous Ammonia Release at Millard Refrigerated Services, Inc., CSB Safety Bulletin, 2015.
[12] A. Brazier, Accident Avoidance, Loss Prevention Bulletin 211, 2010.
[13] Health and Safety Executive, Why leadership is important, http://www.hse.gov.uk/leadership/whyleadership.htm (accessed 23rd March 2020).
[14] Wikipedia, Tlahuelilpan Pipeline Explosion, https://en.wikipedia.org/wiki/Tlahuelilpan_pipeline_explosion (accessed 23 March 2020).

[15] Mexico News Daily, Audit slams Pemex for Inefficient Pipeline Monitoring, Maintenance and Protection, https://mexiconewsdaily.com/news/audit-slams-pemex-for-inefficient-pipeline-monitoring/, 22 February 2019 (accessed 23 March 2020).

[16] Health and Safety Executive, Performance Influencing Factors, http://www.hse.gov.uk/humanfactors/topics/pifs (accessed 23 March 2019).

[17] Health and Safety Executive, Offshore Technology Report OTO 1999 092 Human Factors Assessment of Safety Critical Tasks, (2000).

[18] A. Brazier, Emergency Procedures, Loss Prevention Bulletin 254, 2017.

[19] Anon, Untimely Injection of Foam Into a Warehouse Containing Pesticide Products, Loss Prevention Bulletin 248, 2016.

[20] A. Brazier, Shift Handover, Loss Prevention Bulletin 261, 2018.

[21] International Society of Automation (ISA), ANSI/ISA-101.01-2015, Human Machine Interfaces for Process Automation Systems, (2015).

[22] Engineering Equipment and Materials Users Association, Control Rooms: A guide to Their Specification, Design, Commissioning and Operation, third ed., EEMUA Publication 201, 2019.

[23] Health and Safety Executive, RR446—The Development of a Fatigue/Risk Index for Shiftworkers, http://www.hse.gov.uk/research/rrhtm/rr446.htm (accessed 23 March 2019).

[24] Health and Safety Executive, HSG 256 Management Shiftwork, HSE Health and Safety Guidance, 2009.

[25] Health and Safety Executive, Human Factors: Behavioural Safety Approaches—An Introduction (also known as behaviour modification), http://www.hse.gov.uk/humanfactors/topics/behaviouralintor.htm (accessed 23 March 2020).

[26] Center for Chemical Process Safety of the American Institute of Chemical Engineers, Guidelines for Initiating Events and Independent Protection Layers in Layer of Protection Analysis, CCPS, 2014.

[27] Health and Safety Executive, Human Factors Aspects of Safety Report Assessment, https://www.hse.gov.uk/comah/sram/docs/s12d.pdf. (accessed 23 March 2020).

[28] J. Reason, Organizational Accidents Revisited, CRC Press, 2016.

[29] Energy Institute, Guidance on Human Factors Safety Critical Task Analysis, second ed., (2020).

[30] C. Grounds, L. Miller, Helping humans get it right, in: 14th Global Congress on Process Safety, American Institute of Chemical Engineers, 2018.

[31] A. Brazier, Linking task analysis with other process safety activities, in: Hazards XXIV, Institution of Chemical Engineers, 2014.

[32] Health and Safety Executive, HSG250. Guidance on Permit-To-Work Systems: A Guide for the Petroleum, Chemical and Allied Industries, (2005).

[33] Health and Safety Executive, HSG 253. The safe Isolation of Plant and Equipment, (2006).

[34] Health and Safety Executive, CRR 348/2001. Assessing the Safety of Staffing Arrangements for Process Operations in the Chemical and Allied Industries, (2001).

# Accident investigations—Missed opportunities

## 7.0 Introduction

Investigations into catastrophic events have revealed something of major significance - the key to preventing disasters first lies in ... examining ... near misses and lower-consequence higher frequency occurrences... [1].

A gifted team of Italian engineers once worked on projects over two sites—one near the shore of Lake Como, the other in the slow food capital of Piedmont. The senior instrument engineer had junior staff at both sites, but so much time was spent resolving urgent issues, he was frustrated by how little time was left to impart his wisdom and experience to his juniors. Regularly driving for hours between the two sites, he likened his dilemma to someone with a small blanket on a cold night, he had to make a regular choice between cold feet or a cold neck.

This is the reality of work. There are more things that you should do than you have time to do. *Urgent* often takes precedence over *Important*. External pressures from customers and suppliers, government agencies, head office initiatives, budget constraints, and skill shortages mean that real work involves juggling competing tasks.

It is tempting to approve the temporary trip override, admonish the careless worker for a minor spill, and thank the heavens for a near miss when more serious consequences were avoided by alignment of good fortune.

But these events are gifts to be accepted and unwrapped.

---

**Text Box 7.1**

Failures should be seen as educational experiences. Having paid the tuition fee, we should learn the lessons [2].

---

In this chapter a single story is used to illustrate the ideas of Kletz, the shortcomings he identified in accident and incident investigations,[a] and how they can be improved.

Both Accident and Incident are defined here as "An unplanned event or sequence of events that results in an undesirable consequence" [1].

Accident investigations aim to understand the circumstances leading up to an unforeseen and undesired event, and by uncovering the root causes, prevent repetition.

An effective accident investigation programme includes

1. Recognising which events to investigate
2. Assembling a team with the right mix of skills, knowledge, and experience
3. Collecting information
4. Analysing the information and asking why—continuing past the causal event to root causes
5. Sharing reports and recommendations
6. Implementing proportionate remedial actions and reviewing their effectiveness in practice

---

**Text Box 7.2**

Almost all ... accidents ... need not have occurred. Similar ones have happened before, and accounts of them have been published. Someone knew how to prevent them even if the people on the job at the time did not. Having paid the price of an accident, minor or serious (or narrowly missed), we should use the opportunity to learn from it [2].

---

Major accident investigations may be multiagency and led by an independent chair with access to expert resources. Examples of exemplary accident investigations are the Cullen investigation into the Piper Alpha disaster [3] and the CSB report and video on Texas city. [4]

All of these investigations show that there were previous events, early warnings which were ignored.

It is by properly investigating the less serious events—near misses, losses of containment, process upsets, and minor accidents—and acting swiftly on what is found that the major events are avoided.

This chapter is aimed at those involved in day-to-day operations, whether it be in research, design, construction, production, maintenance, transport, or decommissioning—to everyone working at the sharp end in process or other high hazard industries.

Trevor Kletz identified 10 common failings of accident investigations [2].

---

[a] The Center for Chemical Process Safety (CCPS) definition of accident and incident have some overlap [1].

Incident—An event, or series of events, resulting in one or more undesirable consequences, such as harm to people, damage to the environment, or asset/business losses. Such events include fires, explosions, releases of toxic or otherwise harmful substances, runaway reactions, etc.

Accident—An event that can cause (or has caused) significant harm to workers, the environment, property, and the surrounding community.

Text Box 7.3

1. Accident investigations often find only a single cause
2. Accident investigations are often superficial
3. Accident investigations list human error as a cause
4. Accident reports look for people to blame
5. Accident reports list causes that are difficult or impossible to remove
6. We change procedures rather than designs
7. We may go too far
8. We do not let others learn from our experience
9. We read or receive only overviews
10. We forget the lessons learned and allow the accident to happen again

Let's add one more

11. We examine each accident in isolation—making new recommendations without looking back at how effectively previous recommendations have been implemented.

## 7.1 Accident investigations often find only a single cause

The site manager of a high hazard manufacturing operation spent a good part of the day on the site, walking around, observing, talking, and listening to people. As Trevor Kletz said "… while there (is) room for variation in management style, if (you are) spending less than 3 hours per day on the plant, (you) should ask (yourself) if it (is) enough" [5].

One day, the manager struck up a conversation with an operator standing beside a hose and asked what he was doing (adding water through the drain point on a transfer pump) and why (priming the pump). See Fig. 7.1.

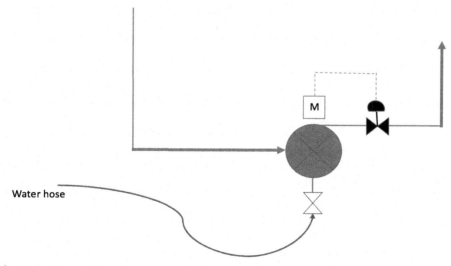

FIG. 7.1 Water hose used to prime pump.

The manager was still pondering when the conversation moved on. They had both worked previously for the same company and strayed onto the subject of deferred pensions. After leaving the area to attend a meeting, the manager was called back an hour later to respond to a serious incident, a spill of several tonnes of flammable solvent from the vent system catchpot. Despite activating an automatic shut-off on all the feeds, the solvent flow continued onto the hardstanding, and a site wide emergency was declared. See Fig. 7.2.

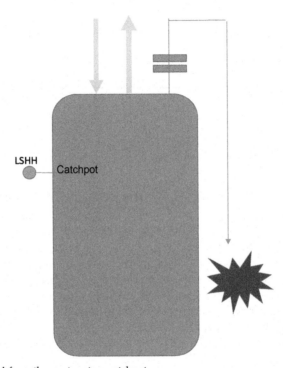

**FIG. 7.2**   Solvent spilled from the vent system catchpot.

The team responded swiftly to the incident. Fortunately, the solvent vapour did not find a source of ignition before the foam blanket was activated. They were lucky. The solvent was collected and disposed of safely, but the loss of containment could have resulted in a serious fire and much worse.

The next day, a manager from another site was brought in to conduct the investigation.

The operator came forward to admit responsibility. After the conversation with his manager, he had moved on to other tasks, forgetting to close the valves on the water manifold or pump drain. Water flowed into the pump suction and back into the condensate vessel above, displacing the solvent which rose into a common vent system ending up in the vent catchpot. An independent high level alarm tripped the feeds as part of an emergency shutdown, but there was no automatic shut-off on the manual water valve. Water continued to flow in and displace the lighter solvent which spilled out through an overflow. See Fig. 7.3.

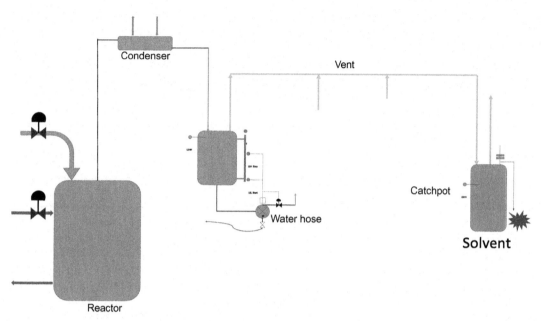

FIG. 7.3  Water from the hose displaced the solvent in the catchpot.

The immediate "cause" of the accident was found to be the carelessness of the operator. He was admonished and sanctioned and he promised to take more care next time.

---

**Text Box 7.4**

Often, accident reports identify only a single cause, though many people, from the chemical and engineering designers down to the last link in the chain, the mechanic who broke the wrong joint or the operator who closed the wrong valve, had an opportunity to prevent the accident. The single cause identified is usually this last link in the chain of events that led to the accident. Just as we are blind to all but one of the octaves in the electromagnetic spectrum, we are blind to many of the opportunities that we have to prevent an accident. But just as we have found ways of making the rest of the spectrum visible, we need to make all the ways of preventing an accident visible [2].

---

A contributory cause was found to be the actions of the manager.

## 7.2 Accident investigations are often superficial

Fortunately, the experienced investigator brought to the site did not stop at the immediate cause. He understood the difference between a causal factor and a root cause.

Causal factor—A major unplanned, unintended contributor to an incident (a negative event or undesirable condition), that if eliminated would have either prevented the occurrence of the incident or reduced its severity or frequency [1].
Root Cause—A fundamental, underlying, system-related reason why an incident occurred that identifies correctable failure(s) in management systems [1].

---

**Text Box 7.5**

Were previous incidents overlooked because the results were, by good fortune, only trivial? The emphasis should shift from blaming the operator to removing opportunities for error or identifying weaknesses in the design and management systems [2].

---

The investigating team started to delve deeper into the task that had puzzled the site manager and caused her to approach the operator in the first place—why had he found it necessary to add water from a hose?

It soon transpired that this was a routine operation, carried out once or twice per shift. The pump which transferred the contents of the condensate vessel as it filled, starting at high level and stopping at low level, often failed to restart. A neat, if undocumented work around had been found by the shift teams. By adding a little water to prime the pump, it then ran without problems. See Fig. 7.4.

This is a good example of the difference between the concept of "work as imagined"—how we think a task will be done—and "work as done" [6], which is how a task changes to adapt to the actual the working conditions—the reality of what operators do to get the job done. This is a key concept we'll come back to in a moment, but for now, back to the pump and the hose.

A proposal was made to replace the water hose with a hard pipe and add an automatic valve to the water inlet. A little clever automation and a slug of water could be added at regular intervals to keep the operation running smoothly. In the event of an emergency shutdown, the water valve would close.

---

**Text Box 7.6**

Even when we find more than one cause, we often find only the immediate causes. We should look beyond them for ways of avoiding the hazards, such as inherently safer design [2].

---

Fortunately, the investigating team rejected this suggestion, asking why the pump needed to be primed with water at all. They did the NPSH calculations (net positive suction head) and

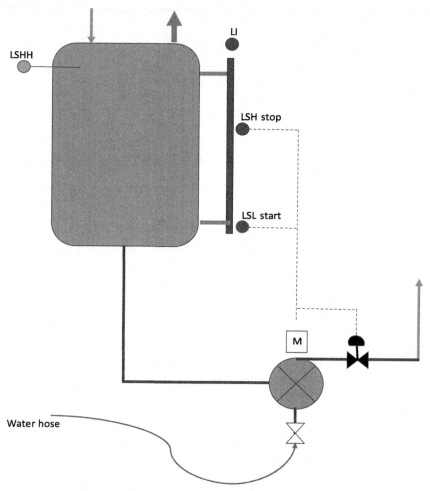

**FIG. 7.4** Pump designed to start and stop on low level, but often became 'vapour locked' due to low NPSH.

discovered that the setting of the low level was too low; the pump was already cavitating before it tripped on low level, leaving it 'vapour locked' and difficult to restart when the level rose again.

Easy! Simply raise the low level trip setting and the problem is solved. No need for water addition at all.

Just one problem. The pump was a large centrifugal pump, designed for continuous operation and high flows, ill-suited for the small intermittent batch duty. It was already a maintenance headache, requiring regular attention. If the proposed changes were to be implemented, the pump would start and stop more often, exacerbating its problems. A proposal was made to replace the pump with a more suitable design.

**Text Box 7.7**

When investigators are asked to look for underlying or root causes, some of them simply call the causes they have found root causes.

One report quoted corrosion as the root cause of equipment failure, but it is an immediate cause. To find the true root causes, we need to ask if corrosion was foreseen during design and if not, why not? Were operating conditions the same as those given to the designer and if not, why not? Was regular examination for corrosion requested, and if so, had it been carried out and were the results acted upon? Senior managers should not accept accident reports that deal only with immediate causes. The causes listed in accident reports sometimes tell us more about the investigators' beliefs and background than about the accidents [2].

## 7.3 Accident investigations list human error as a cause

An accident, by definition, is an unplanned, undesirable event.

In every accident investigation, you will find human error. This is because in every activity, there are human tasks. From the simplest task such as connecting a hose to the drain of a pump, to more complex tasks such as writing, checking, and testing a software program that adds water automatically, you find variation, adaptation, and optimisation.

When nothing bad happens, we see this as normal, how work gets done. It is only when something goes wrong that we call this human error. With the benefit of hindsight, you can work backwards from the undesired event, isolate and point to the human error. But it doesn't mean it is the root cause.

Human beings are not machines who follow rules and execute repetitive tasks perfectly.

What makes human activity so successful is our ability to recognise patterns, spot differences, and adapt to changing environments in a way that rule following machines cannot yet do. And when artificial intelligence and machine learning is advanced enough for machines to take over human tasks (such as driving a car) and something goes wrong, it will be an error in the program or instrument design which will ultimately lead back to a human action.

Trevor Kletz had an experienced practitioner's insight into human error and identified four classes of error [2].

**Text Box 7.8**

Human error is far too vague a term to be useful. We should ask, "What sort of error?" because different sorts of error require different actions if we are going to prevent the errors from happening again

1. A mistake—wrong intention
2. A violation or non-compliance—deliberate breaking of known rules
3. Beyond capability
4. Slip or lapse of attention

Let's examine each of these for the case of the water hose which was left running, causing a loss of containment of a flammable solvent.

---

**Text Box 7.9**

Was the error a mistake—that is, one due to poor training or instructions, so that the intention was wrong. If so, we need to improve the training and instructions and, if possible, simplify the task [2].

---

In this example, the task was not covered by a written instruction. Most routine, apparently low hazard tasks are not covered by detailed operating instructions. And nor should they be. It is almost impossible to write a set of instructions detailed enough to cover every eventuality, but clear enough to be followed sequentially in real time. There may be many volumes written at the time of the original plant commissioning, but chances are they sit on the top of a filing cabinet, covered in dust, or languishing in a computer folder which is only opened during an accident investigation.

---

**Text Box 7.10**

However many instructions we write, we will never foresee everything that might go wrong. Whereas instructions tell us what to do, training gives us the understanding that allows us to handle unforeseen situations [2].

---

The use of hoses was certainly covered by training—how to identify the correct hose material by colour code, how to check integrity and pressure rating, how to loosen and tighten the jubilee clip that attached the hose to the water point, routing the hose to avoid any trip hazard and tidying it away after use.

Whilst it could be argued that the decision to use the hose in the first place was a mistake, it was not an uncommon workaround on that particular plant.

Would the activity had been seen as a mistake if no accident had happened?

Probably not.

---

**Text Box 7.11**

Was the error due to a violation or noncompliance—that is, a deliberate decision not to follow instructions or recognised good practice? If so, we need to explain the reasons for them as we do not live in a society in which people will simply do what they are told. We should, if possible, simplify the task—if an incorrect method is easier than the correct one, it is difficult to persuade everyone to use the correct method—and we should check from time to time to see that instructions are being followed [2].

---

In this example, there was no deliberate violation or noncompliance. The management of change (MOC) process would almost certainly have picked up the danger of uncontrolled water addition, but at the time no MOC form was required for routine production activities that were considered normal to keep the plant running. After the accident, this exception was removed, with the result that the MOC system became briefly swamped with minor, low-risk changes before it was redesigned.

---

**Text Box 7.12**

Was the task beyond the ability of the person asked to do it, perhaps beyond anyone's ability? If so, we need to redesign the task [2].

---

In this example, the task was completely within the capability of the operator.

---

**Text Box 7.13**

Was it a slip or lapse of attention? In contrast to mistakes, the intention may have been correct but it was not fulfilled [2].

---

Although we might argue that the decision to use the hose in the first place was a mistake or a violation, in this example there was clearly lapse of attention (or more precisely a distraction) which led to a slip. Instead of closing the water and drain valves and disconnecting the hose, the operator left the water running. Enough to admonish the operator and tell him to be more careful in future? Or react more firmly and start disciplinary action?

As Trevor Kletz pointed out, exhortations to take more care may close off a recommendation from an accident investigation but is rarely effective in the long term.

---

**Text Box 7.14**

It is no use telling people to be more careful as no one is deliberately careless. We should remove opportunities for error by changing the design or method of working. Designers, supervisors, and managers make errors of all these types though slips and lapses of attention by designers and managers are rare as they usually have time to check their work. Errors by designers produce traps into which operators fall—that is, they produce situations in which slips or lapses of attention, inevitable from time to time, result in accidents. Errors by managers are signposts pointing in the wrong directions [2].

---

Ah, yes. Unintentional management signposts. Should the manager who distracted the operator take the blame?

## 7.4 Accident reports look for people to blame

Blame is a trap we should avoid because it gets in the way of understanding why events unfolded as they did and stops us identifying and fixing the root causes. Avoiding blame is fundamental to accident investigation because it affects the way an investigation is conducted, which leads us to the next common failing of accident investigations—accident reports look for people to blame.

---

### Text Box 7.15

In every walk of life, when things go wrong the default action of many people is to ask who is to blame? The banner headline in my newspaper after a railway accident was "Who is to blame this time?" However, blaming human error for an accident diverts attention from what can be done by better design or methods of operation. To quote James Reason, "We cannot change the human condition but we can change the conditions in which humans work." Even when people ask, "What did we do wrong?" they often find the wrong answer. They find that the instructions were perhaps not clear enough, rewrite them in greater detail and at greater length, and thus reduce the probability that anyone will read them.

To paraphrase G.K. Chesterton, the horrible thing about all the people who work at plants, even the best, is not that they are wicked, not that they are stupid; it is simply that they have got used to it. They do not see the hazards; all they see is the usual people carrying out the usual tasks in the usual place. They do not see the risks; they see only their own place of work [2].

---

When you drive to work, you do not always inspect your vehicle tyres before entering the car. Perhaps you don't keep your hands at the ideal ten-to-two position on the steering wheel. On spotting an unfamiliar road sign, you do not stop and consult the Highway Code before proceeding. You may stray a mile or two over a speed limit in an area free of hazards. You may become distracted by a passenger, a news item on the radio, or a piece of music. Sometimes you arrive at work and have no memory at all of how you got there. The journey to work becomes so familiar that you switch to autopilot.

How we respond when things go wrong matters. Many things often had to go wrong for accidents to happen—the most visible cause, the person turning the valve, is often not the root cause, which lies in the system or process or environment, and blaming people stops us getting to the root causes, or other contributory causes, so we can address and fix them.

So instead of asking who is to blame, or why did they do that, respond to bad news by asking [7]:

- Is everyone ok?
- Is the plant safe, secure, and stable?
- Take me through what you know about how it happened
- What happened just before the event?

Back to the curious case of the water hose.

It was clear that both the operator and the site manager were involved in the accident. But what of the control room operator? Why, the investigator asked, had action not been taken when the first high level alarm sounded?

The control room operator became defensive. He claimed to have reacted swiftly. Several alarms had sounded in quick succession, and it took him a few minutes to accept each one. Although with hindsight it was known that a high level had caused the flurry of other alarms, at the time it was not clear which had happened first. He had gone outside to check the local display on the level device which showed that all was well and then walked downstairs to check that the troublesome pump was running. It was.

It was true, he confessed, that he must have walked past the water hose connected to the drain of the pump without noticing it. The position of the manual globe valve on the water manifold could not be detected visually, and he didn't think to check it was closed. Perhaps he was so used to seeing the hose connected to the difficult pump that he thought nothing of it.

By the time he returned to the control room the high level alarm had reset and he concluded that it had been a false alarm, nothing to be concerned about.

And yet the investigating team knew that at the time the operator claimed to have checked, the levels were continuing to rise and filling the common vent systems.

Another example of human error? Another slip of lapse of attention? Or a cover up?

## 7.5 Accident reports list causes that are difficult or impossible to remove

---

**Text Box 7.16**

Blaming something on human error is as useful as saying falls are due to gravity [8].

---

If an accident report lists human error as a cause, you know that the investigation has not gone deep enough.

The first investigation into the 1986 Chernobyl accident (see Section 4 of Appendix) blamed the shift workers.

> the accident was caused by a remarkable range of human errors and violations of operating rules ....The operators deliberately and in violation of rules withdrew most control and safety rods from the core and switched off some important safety systems [9].

By 1992, the contribution of the RMBK design and the Man Machine Interface was recognised.

> The accident is now seen to have been the result of the concurrence of the following major factors: specific physical characteristics of the reactor; specific design features of the reactor control elements; and the fact that the reactor was brought to a state not specified by procedures or investigated by an independent safety body. Most importantly, the physical characteristics of the reactor made possible its unstable behaviour [10].

Otherwise excellent investigations will detail causes which are equally unhelpful.

---

**Text Box 7.17**

For example, a source of ignition is often listed as the cause of a fire or explosion. But it is impossible on the industrial scale to eliminate all sources of ignition with 100% certainty. Although we try to remove as many as possible, it is more important to prevent the formation of flammable mixtures.

Which is the more dangerous action on a plant that handles flammable liquids: to bring in a box of matches or to bring in a bucket? Many people would say that it is more dangerous to bring in the matches, but nobody would knowingly strike them in the presence of a leak and in a well-run plant leaks are small and infrequent. If a bucket is allowed in, however, it may be used for collecting drips or taking samples. A flammable mixture will be present above the surface of the liquid and may be ignited by a stray source of ignition. Of the two causes of the subsequent fire, the bucket is the easier to avoid. I am not, of course, suggesting that we allowed unrestricted use of matches on our plants, but I do suggest that we keep out open containers as thoroughly as we keep out matches.

Instead of listing causes, we should list the actions needed to prevent a recurrence. This forces people to ask if and how each so-called cause can be prevented in the future [2].

---

It took many years, and intense pressure from international safety organisations, to close down the other RMBK reactors in the Chernobyl power complex. Reactors 1, 2, and 3 continued to operate—despite the known design flaws they shared with Reactor 4—up to the year 2000.

Back to the control room operator and his failure to respond to the alarm.

At the time the control room operator claimed to have checked that all was well, solvent displaced by the water was already filling the vent manifold and flowing back into other vessels.

At first glance it appeared that, at best, he was lazy and incompetent. At worst he was lying.

Quite apart from disciplining the individual, did the control room operators need retraining in the importance of responding to alarms?

Fortunately, the experienced investigator was not so simple minded.

At the time of the accident, the company had only just started looking at new guidelines on Alarm Management [11].

An initial survey found that almost half of all alarms were fleeting—lasting for less than 30 seconds and repetitive—the same alarms flicking in and out. Of the remaining 50%, many of these were status alarms. The operator stopped a pump from the control panel and an alarm sounded to say the pump had stopped, followed by another alarm to signal low pressure, and another to signal low flow. True, but not helpful.

Despite this alarm flooding, the control system records showed that the operator had indeed accepted a high level alarm.

The control system records were analysed further and sure enough they showed exactly what the control room operator had claimed. The level appeared to rise, triggering the first high level switch which activated the pump. The level continued to rise triggering the independent high-high level switch which closed the feeds and stopped the heating. The operator received a flood of alarms.

But then something strange happened, the level started to fall until it triggered the low level switch and the level stabilised.

Level devices are curious things.

In the Buncefield accident (see Section 14 of Appendix) both the Automatic Tank Gauge (ATG) and Independent Ultimate High Level Switch failed to provide protection against overflow. "It is a salutary lesson that the ATG had 'stuck' 14 times in the four months prior to the catastrophic overflow event in 2005" [12]

The continuous level device in the example under discussion was a magnetic float in a tube giving a visual indication with two contact switches to start and stop the pump. As the water displaced the solvent, the float in the tube rose. Water continued to flow from the hose with the manual valve and the solvent was pushed up into the vertical vent pipe. Although the solvent was lighter than water, it soon reached a height of several metres, which was enough to push the float back down until it reached an equilibrium level. See Fig. 7.5.

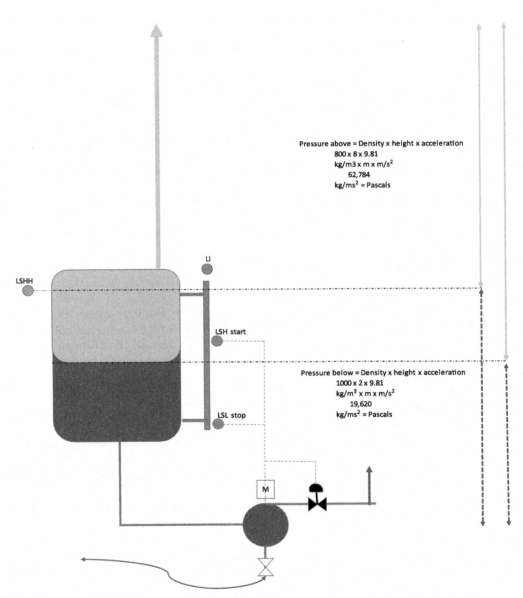

Pressure above = Density x height x acceleration
800 x 8 x 9.81
kg/m3 x m x m/s²
62,784
kg/ms² = Pascals

Pressure below = Density x height x acceleration
1000 x 2 x 9.81
kg/m³ x m x m/s²
19,620
kg/ms² = Pascals

**FIG. 7.5**   As vent filled, high pressure level reset.

The independent high level switch was also a float device. When the high-high level activated, the alarm sounded, closing feeds and shutting off reactor heating. But the mode of failure was the same as the continuous level device, the column of liquid above the float pushed it back down and it reset. Even if it had been a different design, such as a tuning fork style conductivity switch, if it had been calibrated for solvent and not water, the same error might have occurred.

The investigator scratched his head. So, he asked, you restarted the feeds, resumed heating.

No, the control room operator insisted, that happened automatically.

Once again, the truthfulness of the man was called into question. The design of the system was for all safety interlocks to latch. Once a safe state retuned, operations could only restart after a positive decision was made. An active reset.

But when tested, the team discovered a bug in the program. Since a recent upgrade, the interlocks in a particular sector of the plant no longer latched as intended. The interlocks triggered so rarely that no one had noticed. The periodic testing ensured that the interlock functioned, but the reset was not part of the test program.

The wise investigator concluded that there was absolutely no point in retraining the control room operators until the alarm system had been rationalised. And better still, the control room operators should be actively involved in the redesign.

An investigation into the 1984 Bhopal accident (see Section 13 of Appendix) blamed the accident on a deliberate violation by a disgruntled employee.

> The tendency of plant workers to omit facts or distort evidence was also clearly evident after the Bhopal incident...Without a cause, no blame could be established [13].

Attributing ulterior motives to witnesses who present inconvenient evidence can still distort investigations today.

> Start with an open mind, ask the questions and listen to the answers [14].

The rationalisation of the site alarm system was a complex, time consuming, and costly exercise. Had it not been for this apparently minor accident, it might not have started before a far more serious accident occurred. It was just a matter of time.

## 7.6 We change procedures rather than designs

Trevor Kletz urged change through design [2].

---

**Text Box 7.18**

When making recommendations to prevent an accident, our first choice should be to see if we can remove the hazard—the inherently safer approach.

Even if it is impossible at the existing plant, we should note it for the future.

---

If an inherently safer approach cannot be used, consider those approaches higher on the Hierarchy of Controls, for example

---

### Text Box 7.19

Control the hazard with protective equipment, preferably passive equipment, as it does not have to be switched on. As a last (but frequent) resort, we may have to depend on procedures [2].

---

But procedures are at the bottom of the Hierarchy of Controls.

---

### Text Box 7.20

Thus, as a protection against fire, if we cannot use non flammable materials, insulation (passive control) is usually better than water spray turned on automatically (active control), but that is usually better than water spray turned on by people (procedural control).

In some companies, however, the default action is to consider a change in procedures first, sometimes because it is cheaper but more often because it has become a custom and practice carried on unthinkingly [2].

---

When an investigation finds that a "good" procedure did not exist, the obvious conclusion is that the incident would not have happened if one had been available. Hence, the underlying cause is lack of procedures (i.e. there was not a procedure for the task) or procedures are not of sufficient quality (i.e. the procedure for the task was not good enough).

A fundamental problem with people investigating incidents is that they have an over inflated opinion of what procedures can achieve. They tend to assume that procedures can be written for every task, covering every eventuality, and that it is easy to impose rules that say a task cannot be carried out if a procedure is not available.

> But the reality is that real work is complex and unpredictable. We expect (need) people to adapt and use their initiative; working under time pressure and with limited resources. Very often there is simply not the time to find a procedure or physically it is impossible to read a procedure whilst performing the task. The reality is that tasks are performed without reference to procedures most of the time; and most of the time there is no negative outcome. The problem with assuming procedures are more effective than they really are is that every time an incident occurs a new procedure is written, or an existing procedure is expanded. Over time this results in a set of procedures that is unmanageable because there are too many procedures; and the procedures are long and wordy. This adds to the workload to review and update procedures, makes it difficult to find procedures when required and makes them difficult to use in practice.

> The reality is that procedures appear very low on the hierarchy of risk control and will only ever make a fairly modest contribution to safety. Avoiding hindsight bias when considering the role of procedures in incidents can mean that more effective recommendations can be made, leading to a set of procedures that provide effective support to competent people [15].

**Text Box 7.21**

Operators provide the last line of defence against errors by designers and managers. It is a bad strategy to rely on the last line of defence and to neglect the outer ones. Good loss prevention starts far from the top event, in the early stages of design.

Blaming users is a camouflage for poor design [2].

For inherently safer approaches, see Chapter 3.

## 7.7 We may go too far

**Text Box 7.22**

Sometimes after an accident, people go too far and spend time and money on making sure that nothing similar could possibly happen again even though the probability is extremely unlikely. If the accident was a serious one, it may be necessary to do this to reassure employees and the public, but otherwise we should remember that if we goldplate one unit there are fewer resources available to silverplate the others [2].

In the case of the example with the water hose, the investigation concluded that the site Management of Change system needed to expand to cover routine production operations, such as use of a water hoses. Within a few months the system was straining under the weight of this new requirement. Important changes to improve safety were delayed or lost under the avalanche of paperwork required to deal with low-risk operations associated with routine cleaning or plant troubleshooting. When production also started to suffer—as undocumented workarounds were now banned—a long overdue review of the Management of Change system led to a more efficient, risk-based analysis. All change was still screened, but a system of triage allowed a fast turnaround for the low-risk business-as-usual items and concentrated technical resources on the more complex changes.

**Text Box 7.23**

In the United Kingdom the law does not require companies to do everything possible to prevent an accident, only what is reasonably practicable. This legal phrase means that the size of a risk should be compared with the cost of removing it, in money, time, and trouble, and if there is a gross disproportion between them, it is not necessary to remove the risk. In recent years, the regulator, the Health and Safety Executive, has provided detailed advice on the risks that are tolerable and the costs that are considered disproportionate. In most other countries, the law is more rigid and, in theory, expects companies to remove all risks. This, of course, is impossible, but it makes companies reluctant to admit that there is a limit to what they, and society, can afford to spend even to save a life. (If this sounds cold blooded, remember that we are discussing very low probabilities of death where further expenditure will make the probability even lower but is very unlikely to actually prevent any death or even injury.) [2]

## 7.8 We do not let others learn from our experience

Thanks to the influence of Trevor Kletz and others, process safety is always high on the training agenda. In the best companies, information on accidents and incidents are freely and widely shared. But in some companies, there is an increasing tendency for such information to be restricted and tightly controlled, even internally so that people are not aware of the accidents their colleagues have experienced or more importantly the lessons learned.

Trevor Kletz wrote eloquently against this [2].

---

**Text Box 7.24**

Many companies restrict the circulation of incident reports, as they do not want everyone, even everyone in the company, to know that they have blundered. However, this will not prevent the incident from happening again. We should circulate the essential messages widely, in the company and elsewhere, so that others can learn from them, for several reasons as follows:

- Moral: If we have information that might prevent another accident, we have a duty to pass it on.
- Pragmatic: If we tell other organisations about our accidents, they may tell us about theirs.
- Economic: We would like our competitors to spend as much as we do on safety.
- The industry is one: Every accident affects its reputation.

To misquote the well-known words of John Donne: No plant is an Island, entire of itself; every plant is a piece of the Continent, a part of the main. Any plant's loss diminishes us, because we are involved in the Industry: and therefore never send to know for whom the Inquiry sitteth; it sitteth for thee.

---

Whilst individual companies may be less keen to share information about accidents, new organisations and new technologies are helping spread the process safety message. Just a few examples are:

1. The US Chemical Safety Board was set up to "investigate accidents … to identify the … causes so that similar events might be prevented". An independent body, from the time it became operational in 1998 the CSB has issued clear and unbiased reports, empathising with victims and searching for truth that can be used to improve future outcomes rather than apportioning blame. The CSB championed the use of video animations as a powerful tool for sharing the lessons learned through expertly told stories.
2. TV documentaries such as the BBC 2013 *Piper Alpha: Fire in the night* [16] and the long running National Geographic series *Seconds from Disaster* [17] use dramatic recreations of tragic events to chronicle disasters in a memorable way.

FIG. 7.6   Nylon Years, Ramin Abhari

3. Chemical Engineer and artist Ramin Abhari retold the story of the Flixborough disaster in the form of a graphic novel, Nylon Years (Fig. 7.6). [18]
4. The success of films like *Deepwater Horizon* [19] and the recent HBO series *Chernobyl* [20] created by Craig Mazin showed that there is public appetite to understand what went wrong in high profile disasters. Some latitude may have been taken in the dramatisation, but the core of the Chernobyl series is remarkably accurate. In a tense courtroom scene, Jared Harris as Valery Legasov illustrates the science behind the escalating loss of reactor control using clapperboards on a Perspex frame, one of the best explanations—complexity made clear (Fig. 7.7).

FIG. 7.7   Jared Harris in 'Chernobyl', HBO.

5. A student competition run by the Loss Prevention Bulletin (published by Institution of Chemical Engineers) in 2019 [21] resulted in University teams submitting videos, posters, papers, and an original piece of classical music. The winning paper [22] provided two info graphics to support their paper on the Bhopal tragedy (Fig. 7.8).
6. It is hard to remember a time before the internet. Although librarians did a wonderful job, access to information about previous accidents often required you to know what you were looking for. Nowadays Wikipedia is an amazing source of information and powerful search engines can take you all over the world.

But however much more accessible information about major accidents has become, however much broader the media used to share the failures and lessons learned, we still have to make a cognitive leap to draw meaningful parallels with our own situation.

---

**Text Box 7.25**

When information is published, people do not always learn from it. A belief that our problems are different is a common failing [2].

---

## 7.9 We read or receive only overviews

If we only see summaries of accidents, we may think—well that couldn't happen to me, here. Who would be so foolish as to store 40 tonnes of methyl isocyanate in a centre of population and then switch off all the safety systems? If you don't run a nuclear reactor or an offshore oil platform, then what can the reports on these accidents possibly teach?

With almost all major accidents, there are multiple small slips which led up to a major disaster, and for most of us in manufacturing, some of those slips will be unpleasantly familiar. You may think you know what happened in Bhopal in 1984 (Section 13 of Appendix), or Chernobyl in 1986 (Section 4 of Appendix), or Piper Alpha in 1988 (Section 8 of Appendix), but if you read the investigation reports, you may learn something new, something directly useful to your current job.

Even at the level of more minor accidents and incidents in our own workplaces, Trevor Kletz bemoaned the fact that many senior managers failed to challenge the information fed to them.

---

**Text Box 7.26**

Lacking the time to read accident reports in detail, (senior managers) consume pre-digested summaries…, full of generalisations such as there has been an increase in accidents due to inadequate training …The identification of underlying causes can be subjective and is influenced by people's experience, interests, blind spots, and prejudices. Senior managers should read a number of accident reports regularly and, if necessary, discuss them with their authors to see if they agree with the assignment of underlying causes. In any field of study, reliance on secondary sources instead of

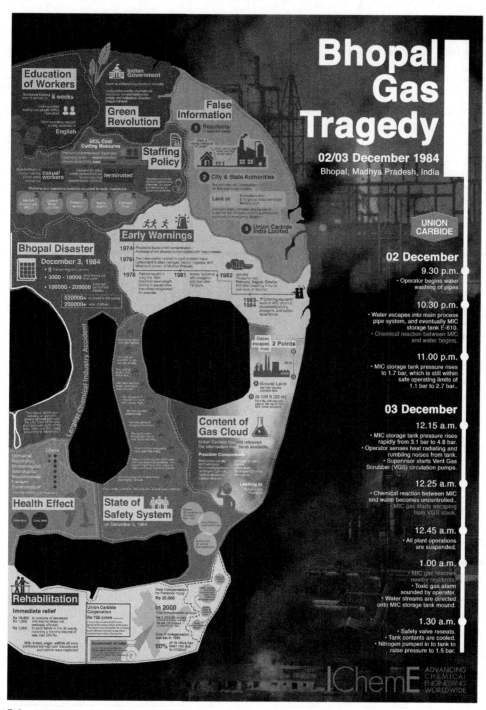

**FIG. 7.8** SIESO medal winning entry 2019.

primary ones can perpetuate errors. Senior managers should be aware that mission or policy statements, though legally required in some countries, have little, if any, effect on safety. People do not change their behaviour as a result of reading a mission statement. They may change as a result of reading an accident report or, better still, taking part in a discussion of an accident. [2]

Far better than reading an accident report in isolation, is a group presentation and discussion of any parallels in the workplace.

Even better if this is a cross-discipline and cross-hierarchy discussion, which provides a platform for wider exchange of information about the real, rather than ideal, world.

> We learn an art by doing that which we wish to do when we have learned it; we become builders by building, and harpers by harping [23].

---

**Text Box 7.27**

Senior managers should also do what they can to stop the spread of the popular view that the consequences of accidents are proportional to the degree of negligence. Similarly, safety is not proportional to the money spent [2].

---

Of course, accident report summaries are important. If we don't hook the reader in at this point, they are unlikely to continue.

But, when it comes to process safety, senior managers have a responsibility to delve deep, to ask questions, to request a complete report and discuss with their teams rather than filing away a pre-digested list of remedial actions. Rather than relying on someone else's interpretation of the lessons learned, senior managers should look for the underlying story.

## 7.10 We forget the lessons learned and allow the accident to happen again

Trevor Kletz said [2].

---

**Text Box 7.28**

Even when we prepare a good report and circulate it widely, all too often it is read, filed, and forgotten.

One method of jerking people out of their familiarity is to show them slides of the hazards they pass every day without noticing them.

On one occasion, I led a discussion of a leak that had occurred from a substandard drain point. Immediately afterward, someone who had been present went into a compressor building that he visited every day. As he walked through the door, he saw a substandard drain point.

---

In the water hose example, sharing details of the accident with other sites in the same company led to identification of other control and trip instruments with common failure modes, other interlocks that didn't latch correctly, and other examples of alarm flooding. But the learning didn't stop there. Other "routine" production tasks were analysed and found to have potential consequences that had not been foreseen. The engineers were kept busy for a while.

---

**Text Box 7.29**

Organisations have no memory. Only people have memories and after a few years they move on, taking their memories with them. Procedures introduced after an accident are allowed to lapse, and some years later the accident happens again, even on the plant where it happened before. If by good fortune the results of an accident are not serious, the lessons are forgotten even more quickly [2].

---

Trevor Kletz suggested some actions to prevent the same accidents from recurring: [2].

---

**Text Box 7.30**

- Include in every instruction, code, and standard a note on the reasons for it and accounts of accidents that would not have occurred if the instruction, procedure, and so on had existed at the time and had been followed. Once we forget the origins of our practices, they become cut flowers; severed from their roots they wither and die.
- Never remove equipment before we know why it was installed. Never abandon a procedure before we know why it was adopted.
- Describe old accidents as well as recent ones, other companies' accidents as well as our own, in safety bulletins and discuss them at safety meetings.
- Follow up at regular intervals to see that the recommendations made after accidents are being followed, in design as well as operations.
- Remember that the first step down the road to an accident occurs when someone turns a blind eye to a missing blind.
- Include important accidents of the past in the training of undergraduates and company employees.
- Keep a folder of old accident reports in every control room. It should be compulsory reading for recruits and others should look through it from time to time.
- Read more books, which tell us what is old, as well as magazines, which tell us what is new.
- We cannot stop downsizing, but we should make sure that the remaining employees at all levels have adequate knowledge and experience.
- Devise better retrieval systems so that we can find details of past accidents in our own and other companies more easily than at present, and the recommendations made afterward.

Everyone forgets the past. A historian of football found that fans would condense the first 100 years of their team's history into two sentences and then describe the last few seasons in painstaking detail. Engineers' poor memories have more serious results.

There is something seriously wrong with our safety education when so many accidents repeat themselves so often.

---

So how do we improve the safety training for engineers?

---

### Text Box 7.31

The first weakness in our safety training is that it is often too theoretical. It starts with principles, codes, and standards. It tells us what we should do and why we should do it and warns us that we may have accidents if we do not follow the advice. If anyone is still reading or listening, it may then go on to describe some of the accidents. We should start by describing accidents and draw the lessons from them for two reasons. First, accidents grab our attention and make us read on, or sit up and listen. Suppose an article describes a management system for the control of plant and process modifications. We probably glance at it and put it aside to read later, and you know what that means. If it is a talk, we may yawn and think, another management system designed by the safety department that the people at the plant will not follow once the novelty wears off. In contrast, if someone describes accidents caused by modifications made without sufficient thought, we are more likely to read on or listen and consider how we might prevent them in the plants under our control. We remember stories about accidents far better than we remember disconnected advice. Whatever the subject, we should build generalities from individual cases; otherwise they have no foundations. The second reason why we should start with accident reports is that the accident tells us what actually happened. You may not agree with my recommendations, but I hope you will not ignore the events I have described. If they could happen at your plant, I hope you will take steps to prevent them, though not necessarily the steps that I have suggested.

A second weakness with our safety training is that it usually consists of talking to people rather than discussing safety training with them. Instead of describing an accident and the recommendations made afterward, outline the story and let the audience question you to find out the rest of the facts, those that they think are important and that they want to know. Then let them say what they think ought to be done to prevent it happening again. More will be remembered and the audience will be more committed than if they were merely told what to do.

Once someone has blown up a plant, they rarely do so again, at least not in the same way. But when he or she leaves, the successor lacks the experience. Discussing accidents is not as effective a learning experience as letting them happen, but it is the best simulation available and it is a lot better than reading a report or listening to a talk.

We should choose for discussion accidents that bring out important messages such as the need to look for underlying causes, the need to control modifications, the need to avoid hazards rather than to control them, the need for inherently safer design, and so on. You can discuss the accidents described in this book, but it would be better to discuss those that occurred in your own plant. The audience cannot then think, we would not do anything as stupid as the people at that plant.

Undergraduate training should include discussion of some accidents, chosen because they illustrate important safety principles. If universities do not provide this sort of training, industry should provide it. In any case, new recruits need training on the specific hazards of the industry [2].

---

It is the success of engineering which holds back the growth of engineering knowledge, and its failures which provide the seeds for its future development [24].

**Text Box 7.32**

It's an easy trap to fall into to think we are learning from incidents when reading about an incident or sharing a safety moment about an incident, however we only truly learn from incidents when we complete actions to change our own systems and processes to prevent similar incidents occurring [2].

## 7.11  We examine each accident in isolation

Let's go right back to the Italian engineer who opened the chapter with his small blanket on a cold night. Did the opening grab your attention? That's what stories do.

His analogy illustrated what every busy person knows. In the real world of work, we continuously make compromises based on the information, time, and resources available. Eric Hollnagel calls this **ETTO**—Efficiency-thoroughness trade-off [6].

We may favour thoroughness over efficiency if safety and quality are the dominant concerns, or efficiency over thoroughness if speed or output is the dominant concern. The ETTO principle states that whilst no activity can expect to succeed without a minimum of either, it is not possible to maximise both efficiency and thoroughness at the same time. The ETTO fallacy is that people are required to be both efficient and thorough at the same time—or rather to be thorough when with hindsight it was wrong to be efficient [25].

Most accident analyses assume that it is possible to reason backwards in time from effect(s) to cause(s) in order to implement specific remedial actions and learn the relevant lessons.

But all too often, the assumptions that guide an investigation (What-You-Look-For) will, to a large extent, determine what lessons are learned (What-You-Find).

If your only tool is hammer, then everything looks like a nail [26].

When something bad happens, there is a tendency for all focus to move to that one event, often at the expense of daily routine and good practice.

If you watch a very junior football match, you will see everyone chasing after the ball. In a well-disciplined senior match, the players will maintain their positions and pass the ball to one another.

Success—in this case safety—comes from both co-ordinated defence and proactive attack.

Whenever a company claims that their first priority is safety, you may think that they are talking nonsense. If safety was really the first priority then all activities would stop, employees would be sent home, and the company closed. No activity is without risk.

However, if we define safety as "the freedom from unacceptable risk" or "the absence of unwanted outcomes" then the claim makes more sense.

"It is usually taken for granted that safety … can be achieved by preventing accidents and incidents … This is a perfectly natural position, since no sane person would hope for something to go wrong or for harm to befall themselves or others. Yet it is also possible to define safety as the presence of acceptable outcomes, and to strive to achieve this by ensuring that things go well rather than by preventing them from going wrong. The terms Safety-I and Safety-II were suggested as a convenient way of referring to these different interpretations. [27]"

If Safety-I involves investigating things that go wrong, then Safety-II focusses on why things more often go well and how we reinforce this. Safety-II is not a replacement for Safety-I, but it asks us to consider activities based on frequency rather than severity of potential consequence alone and to understand what makes work successful and build off that positive capacity to ensure a resilient workplace overall.

Was Trevor Kletz, always ahead of his time, one of the first practitioners of Safety-II? Certainly he insisted on walking around and talking to people, engaging with all levels and disciplines of the organisation in safety conversations, recognising the variability in human performance and promoting inherent safety by design. But Trevor also knew the power of stories.

---

## Text Box 7.33

Case histories grab our attention much more effectively than advice....people may not agree with my advice, but they can hardly ignore the accidents. We should start with the accidents and draw the lessons out of them [2].

---

This chapter opened with a story in the hope of grabbing your attention. It used other stories to illustrate missed opportunities in accident investigations, in the hope that you will remember Kletz's belief that it is only by using stories—case histories of accidents—that we get people to really engage with process safety and keep everyone safe.

# References

[1] Center for Chemical Process Safety, Guidelines for Investigating Process Safety Accidents, AIChemE, Wiley, 2019.
[2] T. Kletz, What Went Wrong? Case Histories of Process Plant Disasters and How They Could Have Been Avoided, fifth ed., Butterworth-Heinemann, 2009.
[3] W.D. Cullen, The Public Inquiry into the Piper Alpha Disaster, HMSO, London, 1990.
[4] Chemical Safety Board, Investigation report - BP America Refinery Explosion. CSB No. 2005-04-I-TX, (2007).
[5] T. Kletz, By Accident....a Life Preventing Them in Industry, PFV Publications, 2000.
[6] E. Hollnagel, The ETTO Principle: Why Things That Go Right Sometimes Go Wrong, Ashgate, 2009.
[7] T. Conklin, Pre-Accident Investigations, Ashgate, 2012.
[8] T. Kletz, An Engineer's View of Human Error, third ed., Institution of Chemical Engineers and Taylor & Francis, New York, 2001.
[9] INSAG-1, Summary report on the post-accident review meeting on the chernobyl accident of the International Atomic Energy Agency's (IAEA's) International Nuclear Safety Advisory Group, (1986).
[10] INSAG-7, The Chernobyl Accident: Updating of INSAG-1, (1992).
[11] EEMUA, 191 Alarm Systems—A Guide to Design, Management and Procurement, EEMUA, 1999.
[12] R. Gowland, Atypical scenarios and dependent failures, Loss Prev. Bull. 230 (2013).
[13] A. Kalelkar, Investigation of large-magnitude incidents: Bhopal as a case study, in: Conference Proceedings, IChemE, 1988.
[14] I. Vince, T. Fishwick, Beware: the witness may be telling the truth!, Loss Prev. Bull. 264 (2018).
[15] A. Brazier, Investigation and bias—procedures, UK, Loss Prev. Bull. 264 (2018).
[16] S. McGinty, Fire in the Night, Documentary, Soda Pictures, 2013. https://www.imdb.com/title/tt2620290/.

[17] Seconds from Disaster, Documentaries, National Geographic Channel, 2004–2012. https://www.imdb.com/title/tt0462133/?ref_=fn_al_tt_1.

[18] R. Abhari, Nylon Years: A Graphic Dramatization of the Flixborough Disaster, CreateSpace Independent Publishing Platform, 2016.

[19] Deepwater Horizon, Film, https://www.imdb.com/title/tt1860357/?ref_=fn_al_tt_1, 2016.

[20] Chernobyl, TV Series, 2019. https://www.imdb.com/title/tt7366338/?ref_=fn_al_tt_1.

[21] SIESO, Medal link, https://www.icheme.org/knowledge/medals-and-prizes/research-and-teaching/sieso-medal/.

[22] SEISO, Medal Winning Entry, https://www.icheme.org/media/12469/sieso-2019.pdf.

[23] Aristotle, The Nicomachean Ethics, Penguin Classics, 2004.

[24] D.I. Blockley, J.R. Henderson, Proc. Inst. Civil Eng. 68 (Part 1) (1980) 719. 47_Y531_Ch38.indd 576 5/21/2009 (2009).

[25] E. Hollnagel, F. Macleod, The imperfections of accident analysis, Loss Prev. Bull. 270 (2019).

[26] Maslow, A Theory of Human Motivation, Wilder Publications, 2013.

[27] E. Hollnagel, Myths and Misunderstandings Safeguard Magazine, (2019).

# Appendix

## A calendar of disasters

| | | |
|---|---|---|
| 1 | January | Feyzin |
| 2 | February | Georgia Sugar |
| 3 | March | Texas City |
| 4 | April | Chernobyl |
| 5 | May | Wanggongchang Armoury Explosion |
| 6 | June | Flixborough |
| 7 | | Chevron Pembroke Explosion |
| 8 | July | Piper Alpha |
| 9 | August | Banqiao |
| 10 | September | Longford |
| 11 | October | Phillips Pasadena |
| 12 | November | Sandoz |
| 13 | December | Bhopal |
| 14 | | Buncefield |

These are brief summaries intended to guide the reader to more comprehensive descriptions of the accident, investigation, and lessons learned.

# 1 January: Feyzin

| Short name | **Feyzin** |
|---|---|
| Date | 4 January 1966 |
| Day and Time | Tuesday 07h00 |
| Place | Feyzin, Lyon, France |
| Short Description | Boiling Liquid Expanding Vapour Explosion (BLEVE) |
| Sector | Energy – Oil Refinery |
| Substances | Propane, LPG |
| Fatalities | 18 |
| Injuries | 81 |
| Company | Elf |
| Longer description | During draining of water prior to routine sampling from the bottom of a 12,000 m3 propane storage sphere, valves froze open releasing thousands of tonnes of liquefied propane. The vapour cloud was ignited by a car about 160 m away.<br>90 minutes later, the sphere ruptured, killing the first responders nearby. Fragments of the ruptured sphere cut through the legs of the next sphere which toppled over. Five of the storage spheres were destroyed. |
| Previous events unheeded | Uncontrolled releases had occurred previously under a butane sphere in August 1964 and under a propane sphere in February 1965. |
| Contributing Factors | Blockage in drain valve caused by ice suddenly cleared and liquid flow could not be stopped<br>Flawed design of sampling operation, poor design of vessel segregation and fire protection, misguided emergency response. |
| Lessons Learned | Avoid draining aqueous liquid or sampling direct from LPG vessels where large volumes of LPG at high pressure could accidently be released.<br>Provide closed drains, sampling, and de-watering pots.<br>Storage spheres can be protected from fire engulfment by better design, fire-resistant insulation, and deluge.<br>Burning hydrocarbon storage vessels are spectacular and unpredictable so emergency response is to evacuate rather than tackle the fire. |
| Key words | Loss of Containment (LOC). Sampling. Emergency response. BLEVE |
| References | Wikipedia accessed 12/04/2020<br>https://en.wikipedia.org/wiki/Feyzin_disaster<br>A Bunn, M Hailwood, Fire and explosion of LPG tanks at Feyzin, France, Loss Prevention Bulletin 251, IChemE (2016)<br>HSE website<br>https://www.hse.gov.uk/comah/sragtech/casefeyzin66.htm<br>ARIA (Analysis, Research and Information on Accidents) database<br>https://www.aria.developpement-durable.gouv.fr/fiche_detaillee/1_en/?lang=en |
| What Trevor Kletz said | Relief valves will not protect a vessel that gets too hot. Pressure vessels must be protected from over temperature as well as overpressure [1]. |

## 2 February: Georgia Sugar

| Short name | Georgia Sugar |
|---|---|
| Date | 7 February 2008 |
| Day and Time | Thursday 19h00 |
| Place | Port Wentworth, Georgia, United States |
| Short Description | Dust explosion at a sugar refinery |
| Sector | Food – Sugar refinery |
| Substances | Sugar |
| Fatalities | 13 |
| Injuries | 42 |
| Company | Imperial Sugar |
| Longer description | An explosion and fire was fuelled by accumulations of combustible sugar dust throughout the packaging building of a sugar refinery.<br>The first dust explosion initiated in the enclosed steel belt conveyor located below the sugar silos, implemented due to food safety reasons. The explosion lifted sugar dust that had accumulated on the floors and elevated horizontal surfaces, propagating much more devastating secondary dust explosions through the building. |
| Previous events unheeded | Between 1980 and 2005, there were 281 explosions in USA involving combustible dust, resulting in 119 deaths and 718 injuries |
| Contributing Factors | Sugar dust accumulation<br>Inappropriate construction materials, poor design and housekeeping, inadequate training or emergency response |
| Lessons Learned | Key role of<br>• design standards<br>• management of change<br>• housekeeping<br>• training and emergency response<br>in safe handling combustible dusts. |
| Key words | Loss of Containment (LOC). Housekeeping. Dust explosion. |
| References | Loss Prevention Bulletin<br>T. Fishwick, The sugar dust explosions and fire at Imperial Sugar Company, Georgia, Loss Prevention Bulletin 266, IChemE (2019)<br>https://www.icheme.org/media/10947/lpb266_pg24.pdf<br>CSB<br>https://www.csb.gov/imperial-sugar-company-dust-explosion-and-fire/<br>Wikipedia<br>https://en.wikipedia.org/wiki/2008_Georgia_sugar_refinery_explosion |
| What Trevor Kletz said | When an accident occurs…outsiders might think that it happened because no one knew how to prevent it. Although the people at the plant at the time, or the designers, may not have known, the information is almost always available somewhere. Few accidents occur because no one knew that there was a hazard [2] |

# 3 March: Texas City

| Short name | Texas City |
|---|---|
| Date | 23 March 2005 |
| Day and Time | Wednesday 13h20 |
| Place | Texas City, Texas, USA |
| Short Description | Vapour cloud explosion |
| Sector | Energy – Oil refinery |
| Substances | Raffinate (a non-aromatic, primarily straight-chain hydrocarbon mixture) |
| Fatalities | 15 |
| Injuries | 180 |
| Company | BP |
| Longer description | Whilst restarting a hydrocarbon isomerisation unit, a distillation tower flooded with hydrocarbons, causing a geyser-like release from the vent stack. Many of the victims were in portacabins (trailers) located nearby. |
| Previous events unheeded | In 1995 the Pennzoil refinery explosion and fire engulfed a portacabin and killed five workers.<br>Between 1994 and 2004, at least eight events reported at the Texas City refinery with flammable vapours emitted by a blowdown drum/vent stack. |
| Contributing Factors | Siting of occupied trailers close to a hazardous unit starting up<br>Corporate culture – cost savings in maintenance, focus on personal safety |
| Lessons Learned | Importance of<br><br>• Understanding, monitoring and reducing process risk<br>• Limiting Occupied Buildings<br>• Replacing Blowdown Stacks<br>• Safety Culture |
| Key words | Loss of Containment (LOC). Occupied Buildings |
| References | *CSB Final Investigation Report on the BP Texas City Refinery Explosion and Fire. ProPublica. January 2007.*<br>https://www.csb.gov/bp-america-refinery-explosion/<br>Loss Prevention Bulletin<br>Lessons from Texas City A Case History Michael P. Broadribb.<br>https://www.icheme.org/media/4611/lpb192pg3-12.pdf<br>Wikipedia<br>https://en.wikipedia.org/wiki/Texas_City_Refinery_explosion |
| What Trevor Kletz said | The "Don't Have" concept can be applied more widely. If chemicals we don't have can't leak, people who aren't there can't be injured or killed. One reason for the large number of deaths and injuries (at Texas City in 2005) was that temporary buildings used by maintenance workers were close to the explosion site. If the buildings had been placed further away from equipment containing hazardous materials — a recommendation that often has been made — the toll would have been lower. Similarly, if no buildings are nearby, they can't be damaged or destroyed by explosions [3]. |

# 4 April: Chernobyl

| Short name | Chernobyl |
| --- | --- |
| Date | 26 April 1986 |
| Day and Time | (Friday night)/Saturday morning 1am |
| Place | Chernobyl, Ukraine, Former Soviet Union |
| Short Description | Fire and explosion with release of radioactive materials |
| Sector | Electricity (Nuclear Energy) |
| Substances | Radioactive isotopes of uranium dioxide, iodine, caesium, strontium, plutonium, and neptunium |
| Fatalities | 31 (immediately after the accident) |
| Injuries | Multiple |
| Long term effects | Chronic illness and economic hardship in Ukraine and Belarus<br>Probably hastened the collapse of the Soviet Union<br>3.5 million Ukrainians claiming state benefits as radiation sufferers (poterpili) in 2000 |
| Environmental Consequences | Deposition of radionuclides across Europe and evacuation of Zone of alienation with radius ~30km around plant. |
| Company | State-owned Nuclear Power |
| Longer description | During a safety test to check back-up power for cooling water, operators lost control of the nuclear reactor. When they attempted to activate an emergency shutdown by inserting control rods into the reactor, the carbon tips of the rods accelerated the reaction, leading to a runaway reaction and explosion. |
| Previous events unheeded | In December 1983 during the commissioning of Ignalina Unit 1 in Lithuania (then part of the USSR) a power surge was observed as the control rods descended into the core. Operators at Chernobyl had not been informed of the causes of this incident or the problems with the carbon tips on the control rods. |
| Contributing Factors | Misguided safety test, which required safety systems to be overridden.<br>Design (inherently unstable – negative void coefficient), Management Culture (Command and Control), Societal culture (Secrecy and Coverup) |
| Lessons Learned | Artificially imposed deadlines lead to shortcuts;<br>Simplified targets in complex environments lead to perverse incentives and unintended consequences;<br>Safety starts with design |
| Key words | Loss of Containment (LOC). Safety test. Trip overrides. |
| References | F. Macleod, Chernobyl 30 Years On, Loss Prevention Bulletin 251, IChemE (2016)<br>Wiki<br>https://en.wikipedia.org/wiki/Chernobyl_disaster<br>INSAG-1, Summary Report on the Post-Accident Review Meeting on the Chernobyl Accident of the International Atomic Energy Agency's (IAEA's) International Nuclear Safety Advisory Group (1986)<br>INSAG-7, The Chernobyl Accident: Updating of INSAG-1 (1992) |
| What Trevor Kletz said | Many companies restrict the circulation of incident reports, as they do not want everyone, even everyone in the company, to know that they have blundered. However, this will not prevent the incident from happening again. We should circulate the essential messages widely, in the company and elsewhere, so that others can learn from them [3]. |

**FIG. A.1**    Photo New Safe Confinement – Chernobyl Power Plant – Fiona Macleod 2018.

# 5 May: Wanggongchang Armoury Explosion

| Short name | Wanggongchang Armory Explosion |
|---|---|
| Date | 30 May 1626 |
| Day and Time | 10h00 Saturday |
| Place | Beijing, China |
| Short Description | Massive explosion |
| Sector | Defence |
| Substances | Gunpowder |
| Fatalities | 20,000 |
| Injuries | Multiple |
| Long term effects | Fall of the Ming dynasty |
| Longer description | Wanggongchang was one of several arms depots manufacturing and storing weapons. To keep armaments secure, the factories were kept within the walls of the densely populated capital.<br>An immense explosion in the armoury killed everyone and obliterated everything within two square kilometres. |
| Previous events unheeded | 220BC – explosion involving the accidental synthesis of gunpowder by Chinese alchemists<br>1552 – seven men were killed and eight injured in London, UK when a spark fell into a container of gunpowder.<br>1560 – Eleven people were killed and 17 more injured when a gun was fired near premises containing gunpowder. Four houses were wrecked, with others damaged.<br>1583 – Three people killed and houses destroyed in a gunpowder store explosion. |
| Contributing Factors | High hazard processing in densely populated area |
| Lessons Learned | Minimise inventory, reduce population close to high hazards |
| Key words | Explosion, Gunpowder |
| References | https://www.historyextra.com/period/tudor/accidental-explosions-gunpowder-in-tudor-and-stuart-london/<br>Wikipedia<br>https://en.wikipedia.org/wiki/Wanggongchang_Explosion |
| What Trevor Kletz said | |

**FIG. A.2**  Photo Forbidden City, Beijing ZhengZhou – Own work, CC BY-SA 4.0, https://commons.wikimedia. org/w/index.php?curid=81925333.

# 6 June: Flixborough

| Short name | Flixborough |
| --- | --- |
| Date | 1st June 1974 |
| Day and Time | Saturday 16h53 |
| Place | Flixborough, North Lincolnshire, England |
| Short Description | Vapour cloud explosion |
| Sector | Chemical – Plastics – Nylon |
| Substances | Cyclohexane |
| Fatalities | 28 |
| Injuries | 36+ |
| Company | Nypro Ltd |
| Longer description | Prior to the explosion a vertical crack in reactor No. 5 was discovered. A bypass was installed between reactors No. 4 and No. 6 so that the plant could continue production. The poorly designed bypass created a large strain on bellows which failed resulting in the escape of a large quantity of cyclohexane. A massive vapour cloud explosion caused extensive damage and started numerous fires on the site. No one escaped from the control room. The fires burned for several days. |
| Previous events unheeded | Energy saving led to agitators being switched off which allowed water to settle out in reactors and contributed to Reactor 5 corrosion and cracking. |
| Contributing Factors | Mechanical failure due to inadequate design of temporary piping. Lack of effective Management of Change, Plant layout, and design. |
| Lessons Learned | Management of Change approvals need expert reviewers. Design codes – importance of following piping standards. Plant layout and control room design Improved regulation – HSAWA 1974 and COMAH 1999 |
| Key words | Loss of Containment (LOC). Vapour Cloud Explosion. Management of Change (MoC). Startup. Cyclohexane. Intrinsic Safety by Design |
| References | HSE http://www.hse.gov.uk/comah/sragtech/caseflixboroug74.htm Health and Safety Executive, 'The Flixborough Disaster: Report of the Court of Inquiry', HMSO, ISBN 0113610750, 1975. R. Turney, Flixborough: Lessons which are still relevant today, Loss Prevention Bulletin 237, (2014) https://www.icheme.org/media/2229/lpb237_p21.pdf Wikipedia https://en.wikipedia.org/wiki/Flixborough_disaster |
| What Trevor Kletz said | The leak was large because only 6% of the hydrocarbon fed to the plant was converted; 94% had to be recovered and repeatedly recycled. The most important recommendation made afterward was that we should look for ways to reduce the amount of hazardous materials in a plant, a process called intensification or minimization. The slogan was: "What you don't have can't leak" [3]. |

# 7 Chevron Pembroke Explosion

| Short name | Chevron Pembroke Amine regeneration unit explosion |
| --- | --- |
| Date | 2nd June 2011 |
| Day and Time | Thursday. 18:00 |
| Place | Pembroke Refinery, Wales, UK |
| Short Description | Atmospheric storage tank exploded whilst being cleaned in preparation for maintenance. |
| Sector | Oil refinery |
| Substances | Amine, diesel, light hydrocarbon (hexane to octane) |
| Fatalities | 4 |
| Injuries | 1 |
| Company | Chevron Limited |
| Longer description | An atmospheric tank was emptied in preparation for maintenance. The contents (amine) were pumped to another tank however a hydrocarbon layer of 0.55 metres depth remained. <br> The Plant Operator relied on a level transmitter that was located 0.59 metres above the base of the tank. <br> Due to the configuration of the tank sump, no flow from the tank drain could not be relied on as an indication it was empty. <br> The contents of the tank were checked by manual dipping but the figure obtained was a significant under estimate. <br> A gas test was taken of the atmosphere in the tank and gave a reading of 67% of Lower Explosion Level (LEL). This was significantly higher than guidelines would allow (typically should be below 10% LEL, and definitely below 25%). <br> Removal of the remaining liquid via vacuum truck involved lowering a non-conducting PVC hose through a manway in the roof. A scaffold pole may have been used to locate the end of the hose at the bottom of the tank and may have allowed static from the hose to create a high energy spark. Spontaneous ignition of pyrophoric substances was another possibility. Neither ignition source could be discounted. |
| Previous events unheeded | Previous incidents on site in 2001 and 2004 resulted in plans to update procedures but these were not properly implemented. |
| Contributing Factors | Lack of understanding of tank design (hydrocarbon accumulation and drain arrangement). Poor gas testing practice. Lack of understanding of risks of potential ignition. <br> Inadequate operating procedures, permit to work, monitoring of work carried out by contractors, and risk assessment. |
| Lessons Learned | Information about hazards must be properly communicated to those carrying out or supervising the work. <br> Gas tests have no value if results are not understood. <br> Safety management systems can degrade. |
| Key words | Tank cleaning, gas test, amine unit, static, pyrophoric. |

| Short name | Chevron Pembroke Amine regeneration unit explosion |
|---|---|
| References | Health and Safety Executive, Chevron Pembroke Amine regeneration unit explosion, 2 June 2011 - An overview of the incident and underlying causes. HSE, 2020 https://www.bbc.co.uk/news/uk-wales-48539681 |
| What Trevor Kletz said | Maintenance is a hazardous activity. Many accidents occur during maintenance, often as a result of poor preparation rather than the maintenance itself [4]. |

**FIG. A.3**    Post-incident – tank 17T302 minus vessel roof. *Contains public sector information published by the Health and Safety Executive and licensed under the Open Government Licence.*

# 8 July: Piper Alpha

| Short name | Piper Alpha |
| --- | --- |
| Date | 6 July 1988 |
| Day and Time | Wednesday evening 22h00 |
| Place | North Sea, Scotland, UK |
| Short Description | Fire and explosion |
| Sector | Oil and Gas, Energy |
| Substances | Hydrocarbon Condensate (mostly Propane), North Sea Crude oil and gas |
| Fatalities | 167 |
| Injuries | Only 61 people survived, many of whom were injured |
| Company | Occidental |
| Longer description | Condensate pump A was isolated for maintenance on its motor drive coupling. When, on the next shift, pump B tripped and could not be restarted, the operators decided to recommission pump A as maintenance had not yet started. However, the pressure relief valve on pump A had also been removed for maintenance under a separate permit. When pump A was restarted condensate leaked from the blind flange which had been installed in place of the pressure relief valve, but not fully tightened up. The escaping gas ignited. The first explosion was quickly followed by an oil pipe rupture and fire. |
| Previous events unheeded | In September 1987, a contract rigger was killed in an accident on Piper Alpha. The accident highlighted the inadequacies of both the permit to work and the shift handover procedures. A golden opportunity to put these right was missed. |
| Contributing Factors | Inadequate communication at shift handover and poor management of permits to work<br>Management failure to confirm that procedures including shift handover and permit to work were effective in practice. Lack of news from the platform was interpreted as meaning there was no problem when in fact it was a sign of poor communication. |
| Lessons Learned | Importance of<br>• Management of change (platform initially only exported oil but was changed to collect and export gas as well)<br>• Personal safety over process safety (fire water pumps were on manual start to protect divers);<br>• Permit to work and shift handover are safety critical communication events and require close attention at all times;<br>• Communication in emergencies is highly complex and unreliable (connected platforms continued to pump oil because Piper Alpha could not contact them);<br>• Emergency response procedures need to be realistic for all possible scenarios (most people died because they followed the procedure that told them to wait at muster until rescued).<br>• As a result of this accident the UK regulation system moved over to a goal based, safety case regime. |

| Short name | Piper Alpha |
|---|---|
| Key words | Loss of Containment (LOC). Permit to Work (PTW). Shift Handover, Management of Change (MoC), Communication |
| References | Piper Alpha [Cullen, W.D.: The Public Inquiry into the Piper Alpha Disaster, HMSO, London, 1990]<br>F Macleod, S Richardson, Piper Alpha – What have we learned? Loss Prevention Bulletin, IChemE 261 (2018)<br>Wikipedia<br>https://en.wikipedia.org/wiki/Piper_Alpha |
| What Trevor Kletz said | The destruction of the Piper Alpha offshore oil platform in 1988 had a similar effect on the UK offshore oil and gas industry as Flixborough had on the chemical industry in the UK in the late 1970s [5].<br>A piece of equipment was brought back into use before maintenance was complete. Senior managers had not noticed that the process was slack, or they had turned a blind eye. Similar incidents continue to occur throughout industry…All management systems and all safety techniques are useless unless the people using them have knowledge, experience and ability [6]. |

**FIG. A.4** Photo Piper Alpha – Bob Fleumer.

## 9 August: Banqiao

| Short name | Banqiao |
|---|---|
| Date | 8 August 1975 |
| Day and Time | Friday 01h00 |
| Place | Zhumadian, Henan Province, China |
| Short Description | Failure of dams during Typhoon Nina |
| Sector | Energy – Hydroelectric |
| Substances | Water |
| Fatalities | Estimated 230,000 drowned immediately with perhaps 1 million dying of hunger and disease. |
| Injuries | Multiple |
| Long term effects | Famine and disease. Disruption to transport and communication. May have hastened the end of the Cultural Revolution. |
| Environmental Consequences | Destruction of farmland |
| Company | State-owned Ministry of Water Resources, China |
| Longer description | After heavy rainfall, the dam on the Banqiao Reservoir breached, releasing a wave of water 10km wide, 7m high travelling at 50km/hour. The "iron dam", a clay structure 24.5m high, was one of 62 dams in Zhumadian that breached in 1975 during Typhoon Nina. |
| Previous events unheeded | Warnings from hydrogeologist Chen Xing |
| Contributing Factors | Typhoon with high rainfall<br>Inadequate Dam design.<br>No Hydrogeological survey.<br>Insufficient sluice gates to release energy. |
| Lessons Learned | Importance of design and maintenance of dams |
| Key words | Loss of Containment (LOC). Stored Energy, Design, Dam Failure |
| References | https://en.wikipedia.org/wiki/Banqiao_Dam<br>https://thechemicaldetective.blog/visit-to-banqiao/ |
| What Trevor Kletz said | Lightning and other so-called Acts of God cannot be avoided but we know they will occur and blaming them is about as helpful as blaming daylight or darkness [5]. |

**FIG. A.5**    Photo Banqiao Dam – Fiona Macleod 2018.

# 10 September: Longford

| Short name | Longford |
|---|---|
| Date | 25 September 1998 |
| Day and Time | Friday 12h26 |
| Place | Longford, Victoria, Australia |
| Short Description | Gas explosion |
| Sector | Energy – Gas |
| Substances | Methane |
| Fatalities | 2 |
| Injuries | 8 |
| Company | ESSO and BHP |
| Longer description | Brittle fracture in a heat exchanger led to a release of gas, fire, and explosion. As well as a tragic loss of life the accident resulted in the loss of gas supply to the local area for two weeks affecting many thousands of people. |
| Previous events unheeded | A similar upset occurred (on 28 August) without failure |
| Contributing Factors | Increase in flow from the Marlin Gas Field and operator response<br>Poor design making process isolation very difficult; inadequate training<br>Excessive alarm and warning systems had caused workers to become desensitised to possible hazardous occurrences;<br>Relocation of plant engineers from the site to offices in Melbourne had reduced interactions with operators and reduced the opportunity they had to detect operational problems;<br>Poor communication between shifts<br>No HAZOP (HAZard and OPerability) analysis of the heat exchange system, which would almost certainly have highlighted the risk of tank rupture caused by sudden temperature change;<br>Company's safety culture oriented to personal rather than process safety |
| Lessons Learned | Information does not equal training (operators had been told about the hazards of low temperature embrittlement of steel but did not understand how it applied to their job).<br>Any organisational change can affect safety critical communication (moving Engineers offsite was a significant factor in this accident).<br>Lack of lost time injuries and good results in safety management system audits are not good indicators of process safety performance. |
| Key words | Gas explosion, design, low temperature embrittlement, alarm management, supervision, HAZOP |
| References | Wikipedia<br>https://en.wikipedia.org/wiki/Esso_Longford_gas_explosion<br>A Hopkins, Lessons from Longford. CCH (2000)<br>D. Dawson, J. Brooks, "The Esso Longford Gas Plant Explosion," report of Royal Commission, State of Victoria, Australia (1999). |

| Short name | Longford |
| --- | --- |
| What Trevor Kletz said | Almost every company has applied the "Don't Have" principle to employees and has reduced their numbers, often successfully. In many cases, though, what firms have called "empowerment" of remaining employees has been a euphemism for loss of support. The classic example was a 1998 explosion at an Esso gas plant in Longford, Australia, that left the whole state of Victoria without natural gas for two weeks. In this case, the company decided to relocate all professional engineers from the plant to headquarters 200 miles away. The official report on the explosion [7] said moving the engineers "appears to have had a lasting impact on operational practices at the Longford plant. The physical isolation of engineers from the plant deprived operations personnel of engineering expertise and knowledge, which previously they gained through interaction and involvement with engineers on site. Moreover, the engineers themselves no longer gained an intimate knowledge of plant activities. The ability to telephone engineers if necessary, or to speak with them during site visits, did not provide the same opportunities for informal exchanges between the two groups, which are often the means of transfer of vital information [3]. |

## 11 October: Phillips Pasadena

| Short name | Phillips Pasadena |
|---|---|
| Date | October 23, 1989 |
| Day and Time | Monday 13h05 |
| Place | Pasadena, Texas, USA |
| Short Description | Release of flammable gas during maintenance leading to vapour cloud explosion |
| Sector | Chemical – Plastics – HDPE |
| Substances | Ethylene, iso-butane |
| Fatalities | 23 |
| Injuries | 314 |
| Company | Phillips 66 |
| Longer description | During regular maintenance operations on a polyethylene reactor a pneumatically operated valve was opened in error. Thirty-nine tonnes of highly flammable gases were released. The vapour cloud came into contact with an ignition source and exploded with the force of 2.4 tons of TNT. |
| Previous events unheeded | Non-compliance with the company isolation standards had not been addressed. |
| Contributing Factors | A valve that should have been isolated was operated to allow process pressure to dislodge a blockage.<br>Control air connections were re-connected the wrong way round which led to the valve going wide open without any further action from the operator.<br>Process design which required regular intervention on a live plant without provision of suitable isolation. |
| Lessons Learned | Importance of<br>• Process hazard analysis<br>• Fail-safe block valves<br>• Inadequate (single) isolation contrary to company standards<br>• Maintenance permitting system<br>• Lockout/tagout procedures<br>• Combustible gas detection and alarm system<br>• Ventilation systems for nearby buildings<br>• Fire protection system maintenance<br>• Distance between high-occupancy structures (control rooms) and hazardous operations<br>• Separation between buildings and equipment |
| Key words | Loss of Containment (LOC), PTW, Maintenance, isolation |
| References | OSHA<br>https://ncsp.tamu.edu/reports/phillips/first%20part.pdf<br>https://en.wikipedia.org/wiki/Phillips_disaster_of_1989<br>K.P. Block, Looking back at the Phillips 66 explosion in Pasadena, Texas: 30 years later. Hydrocarbon processing (2019) https://www.hydrocarbonprocessing.com/magazine/2019/october-2019/special-focus-plant-safety-and-environment/looking-back-at-the-phillips-66-explosion-in-pasadena-texas-30-years-later (Accessed 23 March 2020). |

| Short name | Phillips Pasadena |
|---|---|
| What Trevor Kletz said | The people concerned had no idea of the power of liquids and gases under pressure. If an accident is the result of taking a short cut, it is unlikely that it occurred the first time the short cut was taken. It is more likely that short-cutting has been going on for weeks or months. A good manager would have spotted it and stopped it. If he does not, then when the accident occurs he shares the responsibility for it, legally and morally, even though he is not on the site at the time [3]. |

# 12 November: Sandoz

| Short name | Sandoz warehouse fire |
| --- | --- |
| Date<br>Day and Time | 1 November 1986<br>Midnight |
| Place | Schweizerhalle, Basel, Switzerland |
| Short Description | Firewater applied to the warehouse fire entered local watercourses and killed all aquatic life for a significant distance |
| Sector | Chemical |
| Substances | Agrochemicals |
| Fatalities | None |
| Injuries | 14 Firefighters exposed to smoke from the fire suffered acute health effects of varying severity |
| Long term effects | Decontamination involved a workforce of over 200 and took nearly three months. Thousands of tonnes of contaminated material were removed from the site and surroundings, including the river bed. |
| Environmental Consequences | Severe |
| Company | Sandoz |
| Longer description | A fire occurred in the warehouse. Firefighting appliances used over $10,000\,m^3$ of water at up to an estimated 24,000 litres per minute. The site drainage could not cope with these quantities and flow rates, and so most of the run-off entered the Rhine. Due apparently to confusion amongst the Swiss authorities, the international alarm system for Rhine accidents was only activated after a delay of nearly 24 hours |
| Previous events unheeded | |
| Contributing Factors | Blow torch used for shrink wrapping.<br>Overuse of fire water.<br>Building design was not resistant to fire.<br>Inadequate segregation of chemicals.<br>Drainage system not able to handle fire water. |
| Lessons Learned | Whilst it is important to prevent fire it is also important to plan for dealing with fires when they occur. This includes defining methods to be used and arrangements to contain fire water.<br>Accident results in amendments to the Seveso Directive (Council Directive 88/610/EEC, 24 November 1988) to strengthen requirements for the storage of hazardous substances, |
| Key words | Warehouse, fire water, aquatic life, Seveso |
| References | I. Vince, The Sandoz warehouse fire - 30 years on. Loss Prevention Bulletin 251 (2016) |
| What Trevor Kletz said | "No plant is an Island, entire of itself; every plant is a piece of the Continent, a part of the main. Any plant's loss diminishes us, because we are involved in the Industry…". John Donne quoted by Trevor Kletz [5] |

# 13 December: Bhopal

| Short name | Bhopal |
|---|---|
| Date | 03 December 1984 |
| Day and Time | (Sunday night)/Monday morning 00h05 |
| Place | Bhopal, Madhya Pradesh, India |
| Short Description | Release of toxic gas from a pesticide plant |
| Product | Sevin (pesticide) |
| Substances | Methyl Isocyanate (MIC) |
| Fatalities | At least 3787 |
| Injuries | At least 558,125 |
| Long term effects | Widespread pollution and chronic illness |
| Environmental Consequences | No demolition of the factory or remediation of the land has taken place and widespread contamination of the groundwater in a densely populated area continues unchecked |
| Company | Union Carbide India Ltd |
| Longer description | An uneconomic pesticide plant was being run down prior to closure. Ingress of water into a tank containing 40 tonnes of a highly reactive intermediate led to a runaway reaction and the release of toxic gases into a densely populated neighbourhood. |
| Previous events unheeded | In 1981 a worker was exposed to phosgene and died 72 hours later. In January 1982, a phosgene leak sent 24 workers to hospital. In February 1982, an MIC leak affected 18 workers. In August 1982, a chemical engineer came into contact with liquid MIC, resulting in burns over 30 per cent of his body. In October 1982, there was another MIC leak. A supervisor suffered severe chemical burns and two other workers were severely exposed to the gases. During 1983 and 1984, there were leaks of MIC, chlorine, monomethylamine, phosgene, and carbon tetrachloride, sometimes in combination. |
| Contributing Factors | Accidental entry of water into a tank of highly toxic intermediate Safety culture Removal of safety systems (flare, refrigeration, water curtain), degradation of equipment (level, temperature, pressure instrumentation, and isolation valves), and cost cutting leading to inadequate maintenance and training. |
| Lessons Learned | Uneconomical plants can be particularly hazardous because money is not spent to maintain them and they receive less attention from engineers and technical personnel. Intermediate products can be more hazardous than either the raw materials or end product, but may not receive much attention. Multi-national companies should not export risks by delegating safety responsibility to overseas authorities, especially in developing countries. |
| Key words | Loss of Containment (LOC). Inherent Safety by Design (ISD) |

*Continued*

| Short name | Bhopal |
| --- | --- |
| References | Loss Prevention Bulletin Special Edition<br>https://www.icheme.org/media/2185/lpb240_pg03.pdf<br>Wikipedia<br>https://en.wikipedia.org/wiki/Bhopal_disaster |
| What Trevor Kletz said | "What you don't have can't leak."<br><br>MIC wasn't a raw material or product but an intermediate. Storing it was convenient but nonessential. It could have been used as it was made — then the worst leak would have been a few kilograms from a broken pipe rather than a hundred tons from a tank.<br><br>The "Don't Have" concept can be applied more widely. If chemicals we don't have can't leak, people who aren't there can't be injured or killed. The human toll at Bhopal was so high because a shanty town had grown up near the plant [3]. |

**FIG. A.6**    Bhopal site 31 years after the accident. *Photo Fiona Macleod 2015.*

# 14 Buncefield

| Short name | Buncefield |
|---|---|
| Date | 11 December 2005 |
| Day and Time | Sunday 06h00 |
| Place | Hemel Hempstead, Hertfordshire, England |
| Short Description | Fire and explosion at an oil storage facility |
| Substances | Petrol |
| Fatalities | None |
| Injuries | 43 |
| Longer description | A tank was being filled with petrol. The level gauge was stuck and an independent high-level switch was inoperative. The petrol overflowed through vents at the top, and formed a vapour cloud near ground level, which ignited and exploded. The fires from the explosion then lasted for five days. <br> The independent shut-off switch on the level alarm was not fitted with a critical padlock to allow its check lever to work. <br> Secondary containment (tank bunds) failed and allowed petrol to flow out. Tertiary containment also failed, and fuel and firefighting foam entered groundwater supplies. |
| Previous events unheeded | The level gauge became unreliable after a tank service in August 2005 |
| Contributing Factors | Failures of design and maintenance in both overfill protection systems and liquid containment systems. <br> Process safety controls not maintained to the high standard; <br> Leadership did not manage major hazards effectively; <br> Auditing did not test the quality and effectiveness of systems. |
| Lessons Learned | Overfill protection independent of normal operational monitoring. <br> Adequate monitoring of reliability of safety critical equipment. <br> Verify understanding of how overfill protection instrumentation works before testing and maintenance, and verify effectiveness of overfill protection system. <br> Board-level visibility and promotion of process safety leadership creates positive safety culture in an organisation. |
| Key words | Storage tank overfill protection; process safety management and safety culture, vapour cloud explosion (VCE) |
| References | HSE <br> http://www.hse.gov.uk/comah/buncefield/buncefield-report.pdf <br> Loss Prevention Bulletin |
| What Trevor Kletz said | The group of oil companies that owned the storage depot claimed an explosion of cold gasoline in open air never before had occurred. <br> The people and organizations involved in design, operations and maintenance were unaware of similar explosions in Newark, N.J., in 1983 [8–10], St. Herblain, France, in 1991 [11], Naples, Italy, in 1995 [12], and elsewhere <br> If just one person at Buncefield had known about just one incident, had realized that a similar event could happen there, and had alerted colleagues, the explosion might not have occurred [3]. |

# References

[1] T. Kletz, Protect pressure vessels from fire, Hydrocarb. Process. 56 (1977).

[2] T. Kletz, What Went Wrong? Case Histories of Process Plant Disasters and How They Could Have Been Avoided, 5th Edition, Butterworth-Heinemann, 2009.

[3] T. Kletz, Bhopal leaves a lasting legacy – the disaster taught some hard lessons that the chemical industry still sometimes forgets, in: Chemical Processing, 2009, https://www.chemicalprocessing.com/articles/2009/238/.

[4] T. Kletz, P. Amyotte, A Handbook for Inherently Safer Design, CRC Press, 2010.

[5] T. Kletz, Learning from Accidents, Routledge, 2001.

[6] T. Kletz, By Accident...A Life Preventing Them in Industry, PFV Publications, 2000.

[7] D.M. Dawson, J.B. Brooks, The Esso Longford Gas Plant Explosion, Report of Royal Commission, State of Victoria Australia, 1999.

[8] Report on the Incident at the Texaco Company's Newark Storage Facility, 7th January 1983, Loss Prev. Bull. 57 (1984) 11 Reprinted in Loss Prev. Bulletin, No. 188, p. 10 (Apr. 2006).

[9] M.F. Henry, NFPA's consensus standards at work, Chem. Eng. Prog. 8 (8) (Aug. 1985) 20.

[10] T.A. Kletz, Can cold petrol explode in the open air? Chem. Eng. (1986) 63 Reprinted in Loss Prev. Bulletin, No. 188, p. 9 (Apr. 2006).

[11] J.F. Lechaudet, Assessment of an accidental vapour cloud explosion, Loss Prev. Safety Prom. Proc. Ind. 314 (1995) 377.

[12] G. Russo, M. Maremonti, E. Salzano, V. Tufano, S. Ditali, Vapour cloud explosion in a fuel storage area; a case study, Proc. Safety Environ. Protect. 77 (B6) (1999) 310.

# Index

Note: Page numbers followed by *f* indicate figures, *t* indicate tables, and *b* indicate boxes.

## A

Abbeystead accident, 27
Accident investigation, 248
  reports, 16, 258–261, 258–259*b*, 260*f*
Agriculture, 52, 80
Alarms, 177, 234, 234*b*
  EEMUA 191, 7
  false alarm, 177, 229–230, 258
  management, 7
  nuisance, 234
Alertness, 234–235, 234–235*b*
Allocation of function, 214–215, 214*b*
AL Solutions Inc. accident, 124
American Institute of Chemical Engineers
      9(AIChemE), 8
Ammonia, 17, 28, 66–67, 81, 88, 92, 185
Ammonium nitrate, 157–158
As Low As Reasonably Practicable (ALARP), 13, 40, 55,
      108–110, 207
  control of modifications, 198
  human factors, 207, 242
  inherent safety, 110
  legislation, 109
Audit, 20, 122, 158–159, 163, 174, 200, 221–222
Aviation industry, 58, 185–186

## B

Bayer Crop Science LLC Accident, 165, 188
Behavioural safety, 239–240, 239–240*b*
Belt and braces, 4, 63
Bhopal accident, 79, 81–82, 89–90, 92–93, 112, 261
Blame, 203, 256–258, 257*b*
Blockage, 126, 137–138, 211
Boeing 737 Max accidents, 185–186
Boilers, 163
Bow-tie (diagram), 7, 213, 243
BP Deepwater Horizon/Macondo accident, 188
BP Grangemouth accidents, 187
BP Texas City accident, 5, 82–83, 188
Breakdown maintenance, 119–120
Buncefield accident, 6, 37–38, 57, 96, 255, 259
Bund, 92, 107, 121
Bursting disks (BD), 210–211
Bypass, 58, 120, 150–151, 165, 172, 178–180

## C

Capital expense (CAPEX), 116
Care/Carelessness, 204, 247, 256
Caribbean Petroleum refining accident, 5, 96
Center for Chemical Process Safety (CCPS), 8, 75–76, 241
Challenger Space Shuttle accident, 90
Change.. *See* Modifications
Checklists
  design reviews, 103
  limitations, 16, 229
  maintenance, 126, 145, 149, 161, 165
  shift handover, 229
Chernobyl nuclear power plant accident, 208–209
Chevron Refinery in Pembroke accident, 6, 135,
      141–142
Chevron Richmond refinery accident, 5, 85, 155–156
Chlorine, 17, 92
Chlorofluorocarbons (CFCs), 81
Chronic unease, 223–224
Clapham Junction railway accident, 160
Cleaning/Cleanliness, 100, 117, 124, 139, 156–157,
      263–264
Combined cycle gas turbine (CCGT), 192
Commissioning, 14
Common sense, 3–4, 39, 66–67
Communication, 188, 223, 232
  Esso Longford accident, 187
  human factors, 219, 243
  permit to work, 148, 159, 210
  Piper Alpha accident, 162
  shift handover, 162
Company reputation, 4
Competence, 145, 195–196, 230–232, 230*b*
  assessment, 231
  management systems, 231
  managers, 221
Computer HAZOP (CHAZOP), 41, 112, 185
Conditional modifier (CM), 65
Confined space, 140–141

Construction, 14, 30–31, 99–100, 146, 166, 216–219
    errors, 216–219
Contractors, 115, 144, 166, 215–216, 215b
    HAZOP, 39
    maintenance, 115, 118, 146
    management of change, 188, 198
Control of Major Accident Hazards (COMAH), 109
Control room, 119–120, 177, 234–235
    design, 7, 233–235, 244
    EEMUA 201, 7
Corden Pharmachem accident, 83
Cost-benefit analysis (CBA), 48–49
Cost effective, 1–2
Cost of accidents, 4
Cost of saving a life, 51

**D**
Deepwater Horizon/Macondo accident, 188
Defence in depth, 107–108
Deflagration, 132–133, 212
Degraded safety systems, 120
Dependent failures, 57, 62–63, 68–69
Design
    conceptual, 97–98, 103
    for constructability, 217–218, 217f, 217b
    contractors, 215–216, 215b
    defining, 98
    detailed, 14
    of equipment, 92
    errors, 211–216, 212b, 214b
    front end engineering design (FEED), 15, 98
    human centred, 225
    intent, 216–217, 216–217b
Detonation, 132–133
Dike.. See Bund
Dismantling equipment, 152–153
Dispersion models, 67
Diversity.. See Redundancy and diversity
Documentation.. See also Records
    control, 16, 185
    HAZOP, 10–11, 15, 20, 24, 40
    maintenance, 119, 145
    management of change, 176, 184, 194–195, 197, 199
Double block and bleed, 129–130
Drain/Draining
    Abbeystead accident, 27
    Chevron Pembroke accident, 18
    contamination, 17, 29
    Formosa accident, 207–208
    Fundão dam and Santarém tailing dam accident, 183
    HAZOP of, 17–18, 29
    human factors, 207–208, 243
    Inherent Safety, 92, 103–105t
    maintenance, 129, 150

Management of Change, 191
Dust, 120, 124, 143, 174

**E**
Efficiency-thoroughness trade-off (ETTO), 271
Electrostatic precipitator, 18–19
Emergency de-pressuring (EDP), 177
Emergency instructions, 229–230
Emergency shutdown, 181, 250–251
Energy Institute (EI), 6–7, 77, 243
Engineer, 2
Engineering Equipment and Materials Users
    Association (EEMUA), 7
Engineering Procurement, and Construction
    (EPC), 215
Environmental Harm Index (EHI), 55
Environmental Protection Agency (EPA), 100
Ergonomics, 224
Error.. See Human error
Esso Longford accident, 19, 94–95, 187
Ethers, 88
Ethylene, 17, 132–133, 186–187, 222
    plant, 186–187
Ethylene oxide, 17, 93
European Process Safety Centre (EPSC), 8
Evangelos Florakis Naval Base accident, 179–180

**F**
Failure Modes and Effects Analysis (FMEA), 197–198
False alarm, 177, 229–230, 258
Fatigue (human)
    index, 236
    management of, 235–236, 235b
Fault tree analysis (FTA), 59
Finistère, accident, 210
Firefighters, 97, 179, 229–230
Firefighting water, 49–50
Flash fire, 212
Flixborough accident, 16, 52, 75, 79–81, 92–93, 178–179,
    178f, 186, 249, 265f
Fluid Catalytic Cracking Unit (FCCU), 136
Flushing, 87–88, 132–133, 135–136
Fool-proof, 130, 139f, 155
Foreign Object, 155–156
Formosa accident, 5, 207–208
Fractional dead time (fdt), 56–58
Friendly systems, 75
Front End Engineering Design (FEED), 15, 98
FTA. See Fault tree analysis (FTA)
Fundão and Santarém tailing dam accident, 183

**G**
Gas test, 133, 138, 141–142
Grinding, 132–133, 135–136

**H**

Handover, 147, 162, 226, 229, 232
Hazan.. *See* Quantified Risk Assessment (QRA)
Hazard and Operability (HAZOP), 10
    audit, 20
    automated, 40–41
    batch HAZOP, 41
    check-lists, 16
    deviation, 10–11
    guide to best practice, 12
    human factors, 34
    human/procedures HAZOP, 41
    limitations, 30–38
    method, 12
    organization of, 20–30
    parameters and guidewords, 11–12
    recording the output from, 12–13
    repeat, 15
    retrospective, 15
    reviewing previous reports, 15–16
    safeguards, 33–35
    scribe, 10–12, 22, 24
    team leader, 12, 20, 23–24
    team members, 21–23
    timing, 13–14
Hazard Identification (HAZID), 15, 24, 101–102,
        197–198
Hickson and Welch Ltd. accident, 187
Hierarchy of controls, 107
High-density polyethylene (HDPE), 126
High Integrity Pressure Protection Systems
        (HIPPS), 110
High Temperature Hydrogen Attack (HTHA),
        28, 91
Hindsight, 208, 254, 262–263
Hoses, 29, 99, 111, 174
Housekeeping, 143, 162
Human centred design, 225
Human error, 33, 254–256, 254–256*b*
    construction, 216–219
    design, 211–216
    maintenance, 209–211
    management, 209, 220–224
    operator, 205–206
    probability of, 224–240
    qualitative analysis, 242–244
    quantitative analysis, 240–242
    recovery, 207, 208*t*
Human Error Probability (HEP) data, 241
Human factors, 6–7, 34, 115, 144, 204, 213, 221, 224, 236,
        243
    allocation of function, 214–215
    analysis (quantitative and qualitative), 240–244
    competence in, 244

Performance Influencing Factors (PIF), 225–230, 241,
        243
Human factors engineering (HFE), 7, 213, 218
Human Machine Interfaces (HMI), 233, 233*b*
Human/procedures HAZOP, 41
Husky Energy Refinery accident, 5, 136
Hydrochlorofluorocarbons (HCFCs), 81
Hydrofluoric acid (HF), 174
Hydrogen, 182*f*
Hydrogen sulphide, 145


**I**

Imperial Chemicals Industries (ICI), 6, 9, 14, 115,
        118–119, 222–223
Incident investigation.. *See* Accident investigation
Independent protection layers (IPLs), 63
Inherent safety, 2–4, 73
    attenuation/moderation, 91–92, 91*f*
    definitions, 75–77
    elimination, 95–96
    Inherently Safer Design (ISD), 76
    Inherently Safer Technology (IST), 76
    intensification/minimisation, 85–88, 86*f*, 87*t*
    limitations of effects, 92–93
    principles of, 85–97
    segregation, 97
    simplicity/simplification, 93, 94*f*
    substitution, 88–91, 88*f*
Inhibits and overrides, 93, 110, 120, 163, 173, 180,
        208–209
Inspections, 85, 91, 117, 121–123, 141, 146, 216, 218,
        239
Institution of Chemical Engineers (IChemE), 6, 12, 22
Instruments, 177
    safety instrumented system (SIS), 7, 34, 63, 120, 181,
        210
Interlocks, 110, 137, 165, 172, 208, 261
International Association of Oil &
        Gas Producers (IOGP), 7
Intuitive design, 206–207, 207*f*
Investigation.. *See* Accident investigation
Iron sulphide, 115
Isolation, 5, 18–19, 37, 87*f*, 126–132, 150, 159
    certificate, 162
    double block and bleed, 129–130
    electrical, 130–131
    Lock Out Tag Out (LOTO), 131
    positive, 99–101, 128–129, 165
    process, 127–130, 244
    proved, 130
    securing, 131
    self, 118
    valve, 126–127, 129–130, 163, 172

**J**

Job Safety Analysis (JSA), 197–198

**K**

Ketone-aldehyde (KA) mixture, 80–81
Key Performance Indicators (KPI), 7, 123

**L**

Labels, 238–239, 238*b*
Latent faults, 56–59, 209, 216, 241
Layers of Protection Analysis (LOPA), 37, 63–66, 241
Lessons
    forgetting, 268
    learn the, 4, 247, 264–266, 268
    from the past, 38, 272
Lifecycle, 112
Liquefied Petroleum Gas (LPG), 135
Live-line electrical work, 117
Locks, 161
    Lock Out Tag Out (LOTO), 126, 131
    multi-hasp, 131
    on valves, 118, 131, 165
Loss prevention, 1
Lost-time accidents, 50, 221
Lower explosion limit (LEL), 138, 141–142

**M**

Maintenance
    assessment and method statement, 144–145
    breakdown maintenance, 119–120
    diagnosing faults, 124–125
    dismantling equipment, 152–153
    errors, 209–211
    faults that develop gradually, 119–120
    inspections (of site during maintenance), 216
    live systems, 126–127
    monitoring (daily checks and inspections), 158–159
    preventative, 121–124
    repair and overhaul, 156–157
    signs and labels, 239
Major Accident, 101, 105, 119, 187–188, 221, 224, 248, 266
    Control of Major Accident Hazards (COMAH), 109
    Major Accident to the Environment (MATTE), 55
Management (managers)
    competence, 221
    errors, 220–224, 221*b*
Management of Change (MoC), 193
Man Machine Interface, 258
Manoeuvring Characteristics Augmentation System (MCAS), 185–186
Marcus Oil and Chemical site accident, 156
Mary Kay O'Connor Process Safety Center, 7–8, 127
Metal industry, 119

Methyl isocyanate, 81
Mexico City accident, 82
Millard Refrigerated Services accident, 5, 28, 219
Mobile Elevating Work Platform (MEWP), 214
Modifications
    creeping change, 16, 172, 191
    temporary, 40, 131, 178–180, 193–194

**N**

Nitrogen, 141, 155, 166
North Tees Works, 118
Nuclear industry, 51, 58, 69, 112, 116, 208–209

**O**

'One size fits all' approach, 228–229
Operator (process technician), 120, 205–209, 205–206*b*, 214, 219, 223–224, 249
    Bayer Crop Science accident, 165
    blame, 203, 263
    Chevron Pembroke accident, 135
    control of modifications, 174
    control room, 195, 233, 257–258
    Corden Pharmachem accident, 83
    error rate, 58, 161
    Esso Longford accident, 19
    and HAZOP, 29
    Piper Alpha accident, 232
    start-up, 180, 186
Organisational change, 186–190
Organisations have no memory, 4, 269
Overrides.. *See* Inhibits
Oxygen, 238–239, 238–239*b*

**P**

Pemex pipeline accident, 221–222, 222*b*
Performance influencing factors (PIF), 225–230, 241, 243
Permit-to-work, 5, 118, 127, 132, 144, 147–148, 166, 195, 210, 232
    auditing, 158–159
    electronic, 148
Phillips Petroleum chemical plant accident, 7, 126
Pigging (receivers/launchers), 137
Pipeline, 132–133, 137, 150, 217, 221–222, 237
Piper Alpha, 37, 82
Piping and Instrumentation Diagram (P&ID), 12
Point of work risk assessments, 149–150
Polyvinyl chloride (PVC), 207–208
Pont de Buis, accident, 210
Potential Loss of Life (PLL), 54
Pressure relief devices, 210
Pressure safety valves (PSV), 157, 210–211
Pressure systems, 121

Pre-start-up safety review (PSSR), 165, 199, 216
Preventative maintenance, 121–124
   monitoring backlog, 123
   schedules, 122–123
Probability of failure on demand (PFD), 63
Procedures
   document control, 185
   maintenance, 145
   operation, 94
   start-up and shutdown, 101
Process Hazard Analysis (PHA), 9, 100, 212
Process Hazard Review (PHR), 15
Process safety, 1, 9, 50, 144, 221
Programming errors, 219–220
Protective equipment, 121–122
Protective systems, 121
Purging, 100, 138–139, 166
Pyrophoric, 115, 136, 141–142

**Q**

Qualitative analysis, 10, 47, 109
   human factors, 241–244
Quantified Risk Assessment (QRA), 240–242
   conditional modifier (CM), 65
   consequence analysis, 66–68
   cost-benefit calculations, 52
   dependent failures, 57
   dispersion models, 67
   fault trees, 59–63, 60–61*f*
   frequency/probability assessment, 56–66
   latent faults, 56–59
   risk criteria, 51–56

**R**

Reasonably practicable, 16, 40, 48, 100, 263–264.
     *See also* As Low As Reasonably Practicable
     (ALARP)
Records, 11–13, 199–200, 211. *See also* Documentation
Redundancy and diversity, 63, 69*b*, 90*b*
Refrigerant, 81
Reliability Based Inspection (RBI), 122
Reliability Centred Maintenance (RCM), 122
Relief valves, 52, 56–58
Reputation
   company, 4
   industry, 264
Retrospective HAZOP, 15
Returning to service.. *See* Start-up
Reverse flow, 17
Risk analysis, 47
Risk assessment and method statement (RAMS),
    144–145
Risk Management Plans (RMP), 100

**S**

Safety case, 15–16
Safety critical devices, 210–211
Safety Critical Task Analysis (SCTA), 7, 242–243
Safety-I and Safety-II, 272
Safety instrumented function (SIF), 63
Safety instrumented system (SIS), 34, 63, 71, 103, 120, 181,
    210
   degraded, 59, 120
   IEC 61508/61511, 63, 71
   trips, 21, 34, 37, 52, 57, 62, 95, 123–124, 163,
     180–181, 253
Safety Integrity Level (SIL), 63
Safety record, 221
Safety report, 15–16
Safety systems
   degraded, 120
Scaled Risk Integral (SRI), 54
Shift handover.. *See* Handover
Shutdown, 117–118, 125–126, 132, 136, 159
   check-list, 229
   HAZOP of, 20, 30, 32
   and Inherent Safety, 101
Signs, 209–210, 238–239, 238*b*
Software, 16, 41, 185–186, 219
   control, 21, 28, 35
Spare equipment (process), 120, 133, 207
Spare parts, 153, 158, 175–177
Staffing arrangements, 188, 213, 236–238, 237*b*
Start-up, 65, 95, 124, 163–166, 229
   control of modifications, 180–181
   HAZOP of, 20, 32
   and Inherent Safety, 101
   leaks, 163
   Pre-Start-Up Safety Review (PSSR), 165, 199
   Texas City accident, 82–83
Static electricity, 84, 84*f*
Sterigenics accident, 5, 93
Storage
   Buncefield accident, 37, 96
   Evangelos Florakis accident, 179
   HAZOP of, 30
   and Inherent Safety, 91–92, 95, 99
   intermediate, 41–42, 81, 86
   tanks, 11, 37, 110, 146, 166
   Tianjin accident, 97
Stored energy, 142, 149, 153*f*, 164, 206
Stores, 177, 195
Supervisor, 118, 147, 150, 160, 188, 210, 237

**T**

Tankers, 96, 101, 111, 174, 206, 228, 238
Tanks, 17, 28–29, 37, 100, 135, 140, 177. *See also* Storage

Task Improvement Process (TIP), 243
Technical errors, 212
Technician (maintenance technician), 132, 160, 177, 204–205, 210
Temporary building, 82–83, 96
Test interval, 57–58
Texaco Pembroke Refinery accident, 18, 18*f*
Tianjin accident, 97
Toolbox talks, 148–149
Tosco Avon Refinery accident, 123
Training, 230–232, 230*b*
    courses, 231–232
    *vs.* competence, 230–231
'Trial and error,', 124
Trips, 95, 163, 253. *See also* Safety instrumented system (SIS)
    Buncefield accident, 37
    dependent failures, 62
    HAZOP, 21, 34
    overriding, 124, 180–181
    spurious, 58
    testing, 57, 123–124
    *vs.* relief valves, 52
Turnaround/outage shutdowns, 126

**U**
UK Chemical and Downstream Oil Industry Forum (CDOIF), 56
UK Health and Safety Executive (HSE), 5–6, 54, 67, 76–77, 97, 109, 147, 159, 241, 263–264

Railway Inspectorate, 48
UK National Health Service (NHS), 237–238
US Center for Chemical Process Safety (CCPS), 75–76
US Chemical Safety and Hazard Investigation Board (CSB), 5, 115, 264
'User-friendly design,', 75
US Ink/Sun Chemical Corporation accident, 212

**V**
Valve and Instrument Criticality Analysis (VICA), 214
Vapour-phase reactors, 87
Vinyl chloride monomer (VCM), 191
Violation, 161, 254–256, 258. *See also* Workaround
Voting systems, 58–59

**W**
Warehouse, 30, 49–50, 229–230
Welding, 88, 133, 156
Workaround, 149, 242–243, 255, 263–264
Working environment, 140, 196, 233
Written instructions, 225–227, 225–227*b*
    emergency instructions, 229–230
    format of, 228–229, 229*b*
    fundamental issues with, 227, 227*b*
    general philosophy, 228
    limitations, 227–228

Printed in the United States
By Bookmasters